Power Management for Internet of Everything

EDITORS

Mathieu Coustans
Catherine Dehollain

Ecole Polytechnique Fédérale
de Lausanne (EPFL), Switzerland

Tutorials in Circuits and Systems

For a list of other books in this series, visit www.riverpublishers.com

Series Editors

Peter (Yong) Lian

President IEEE
Circuits and Systems Society
York University, Canada

Franco Maloberti

Past President IEEE
Circuits and Systems Society
University of Pavia, Italy

Published, sold and distributed by:
River Publishers
Niels Jernes Vej 10
9220 Aalborg Ø Denmark

River Publishers
Lange Geer 44
2611 PW Delft
The Netherlands

Tel.: +45369953197 www.riverpublishers.com

Availability: June 2018

ISBN: Print: 978-87-93609-83-9
     E-book: 978-87-93609-82-2

Library of Congress Cataloging-in-Publication Data: June 2018
Editors: Mathieu Coustans, EPFL, Switzerland and Catherine Dehollain, EPFL, Switzerland
Title: Power Management for Internet of Everything

Table of contents

Introduction

Internet of everything or internet of things on itself could be approached from many angles. Considering the internet of thing as an extensive network of connected things, it can be computing devices, digital entities, mechanical machines, objects, or even people, having the ability to collect and exchange data over a network without needing a human interaction. While internet of everything, include a human compound. When a network is considered, it implies an infrastructure which consumes energy. Therefore, this critical aspect linked to energy has to be considered at the system level.

Managing power for the Internet of Everything (IoE) is a challenging task because the devices must always be powered up and can be located anywhere, including harsh, remote environments. It is often impossible to run a wire to a device. CMOS technology is at the heart of many recent developments in the design of integrated circuits. Moore's Law has served as the guiding principle for the semiconductor industry for several years. This trend is still moving forward as the state-of-the-art sub-nm scaled CMOS technologies, for applications ranging from high-performance computing down to ultra-low- power mobile applications are developing. Circuit and system designers all around the globe are leveraging the large device density and processing power of modern technology toward new applications for more smart and interconnected world.

This book targets post-graduate students and engineers working in the industry, willing to understand and connect system design with the system on chip (SoC) and mixed-signal design as abroader set of circuits and systems. The field of IoT, spanning from data converters for sensor interfaces to radio and software applications, is covered with emphasis on power and energy management. This book tries to target a right balance between academia and industrial contributions.

Several advanced topics in the area of Power Management, Analog and Mixed-Signal Circuits and Systems are explored by reviewing in this book their fundamental principles. The first chapter is linked to the personalized medicine domain by covering bio-sensors co-integration with nano-technology and CMOS circuits. Having seen power assets and challenges of such a technology for the circuit and system designer, the remote powering and some sensors solutions are reviewed in the second chapter. The third chapter covers a first industrial contribution dedicated tothe remote powering of a wireless sensor network through the RF field. The concept of very low current consumption (in the order of μA or nA)and their transient behavior are also covered in this third chapter. The fourth chapter is dedicated to the measurement and the analysis of dynamic current profiles for low power consumption. The digital and large-scale integrated circuits are covered by an academic contribution in chapter five and by an industrial contribution in chapter seven. Hall sensors dedicated to automotive applications constitute Chapter six.Chapter eight is dedicated to the design of CMOS oscillators for timers by considering the duty-cycling of the active mode. Finally, system-level power management (including the cloud) are discussed in the last chapters of the book.

MATHIEU COUSTANS AND CATHERINE DEHOLLAIN

CHAPTER 01

Applications in Biosensing of Power Delivery

Sandro Carrara

Ecole Polytechnique Fédérale
de Lausanne (EPFL), Switzerland

This chapter is about the application of power delivery in biosensing. In particular, remotely-powered implantable-systems capable to provide real-time biosensing will be introduced and deeply discussed.

1 Chips under the skin?

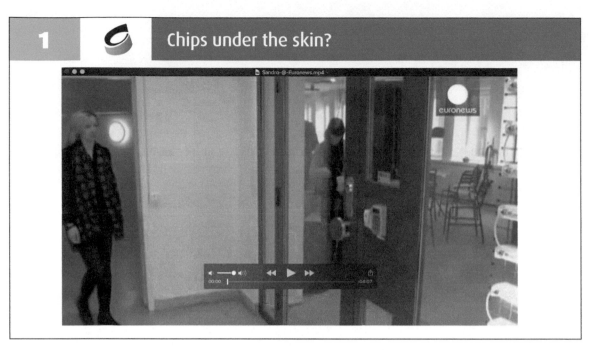

As an introduction to the topic, a video is available on in the net, where several applications of active or passive electronics chip implanted under the skin as shown and their applications discussed.
(copy/past in a browser the following link:
http://www.euronews.com/2015/06/23/rfid-chips-a-key-to-more-or-less-freedom)

2 CommentFully-Connected Human++

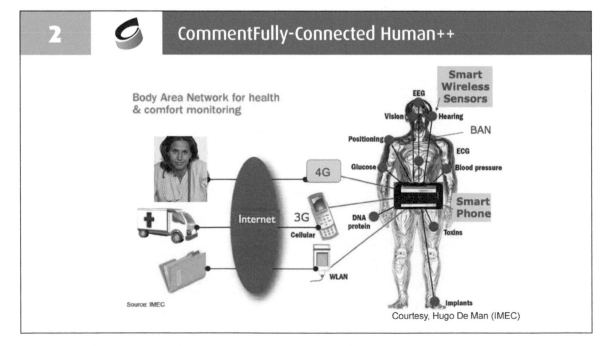

Courtesy, Hugo De Man (IMEC)

Of course, the trend in the development of such a implantable devices as well as their introduction into the market is driven by the idea to foster the monitoring of patients as well as healthy people in their daily-life at home too in order to provide prevention to many diseases or personalised therapy on chronic diseases.

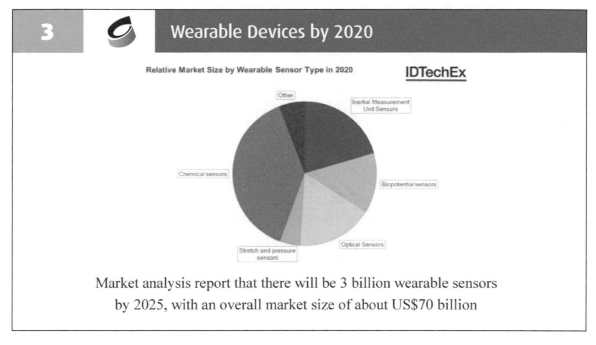

Market analysis report that there will be 3 billion wearable sensors
by 2025, with an overall market size of about US$70 billion

Forecasts from market analysis say that we are facing a period of huge expansion of this area of research and development till reaching 3 billion wearable devices by 2025 for a total value of about 70 billion in USD.

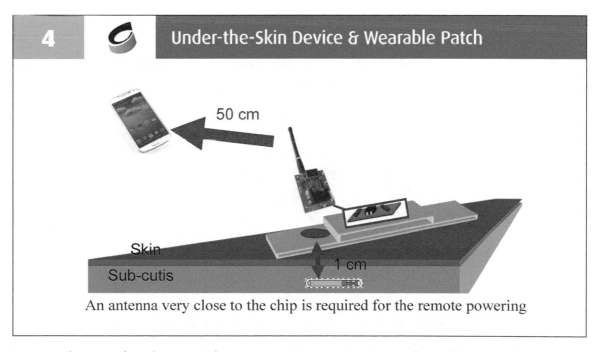

An antenna very close to the chip is required for the remote powering

Having this in mind, we have started years ago in EPFL to investigate actual possibilities to realise such new technologies, even with the idea to test the real limits of such an approach. The core-idea was to develop a so-tiny implant to support insertion by a siring (so, no surgery) while supporting the detection of several diseases biomarkers at molecular level and be connected to our smart phones by means of a wearable patch.

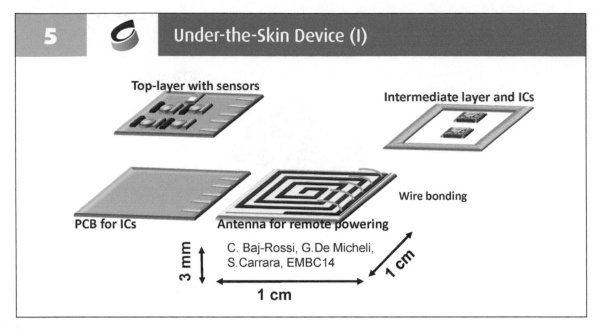

Of course, the critical part of the development was the design of the implantable device, for which we decided to investigate two different options. In a first approach, the implantable device was conceived as multi-layer structure including a coil to receive the power and transmit data, a PCB hosting all the required integrated circuits (IC), a silicon platform hosting several electrochemical biosensors. Wire bonding was chosen to provide electrical contacts.

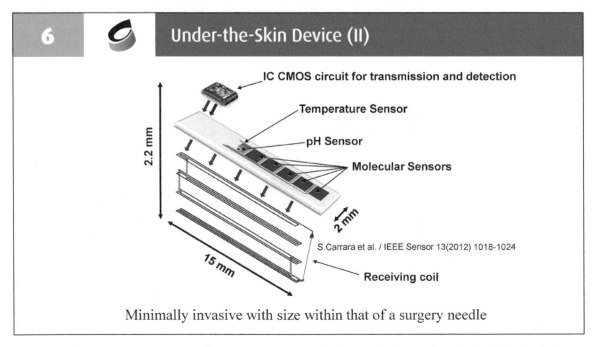

A second design was instead conceived in a way to integrate all the biosensors, the IC, and the receiving coil directly a on silicon substrate.

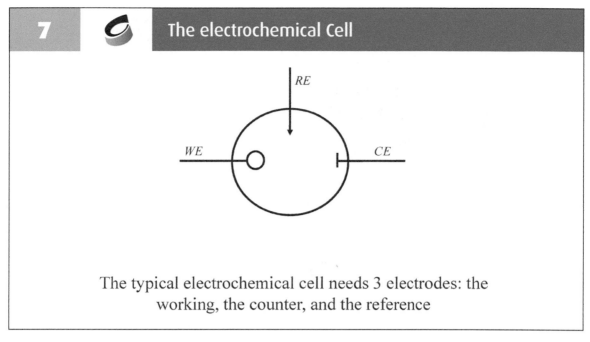

7 ⏣ The electrochemical Cell

The typical electrochemical cell needs 3 electrodes: the
working, the counter, and the reference

In order to realised the electrochemical biosensors needed to address the detection of diseases biomarkers at molecular level, electrochemical cells have to be fabricated and correctly driven by designing the typical three-electrodes: the working electrode (WE), the Counter (CE), and the Reference (RE) ones.

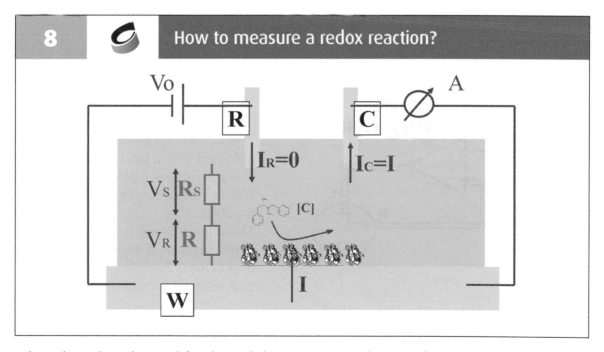

8 ⏣ How to measure a redox reaction?

These three electrodes are definitely needed in amperometric detection of a specific molecule since we need to measure a current by collecting at the CE the electrons emerging from the redox reactions occurring at the WE, meanwhile supporting the reactions by precisely controlling the potential at the electrochemical interface. To assure that, the right potential in then applied at the RE without circulating any current.

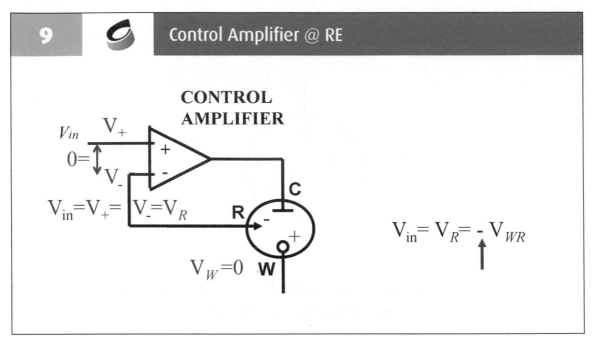

To apply the right potential at the RE without circulating any current, we can use an operational amplifier (OpAmp) by exploiting one of the usual input (e.g., the "-" input) to transfer at the CE a potential applied to the other input (e.g., "V_{in}" at the "+' input). This works thank to the high input impedance we usually have in OpAmps.

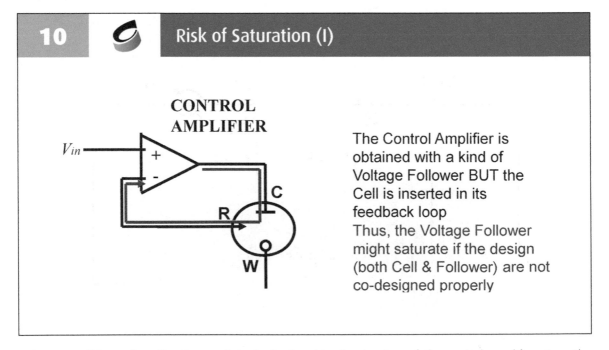

However, this configuration has a draw back: the risk of saturation of the control amplifier since the electrochemical cell is now located inside the closed-loop of the amplifier.

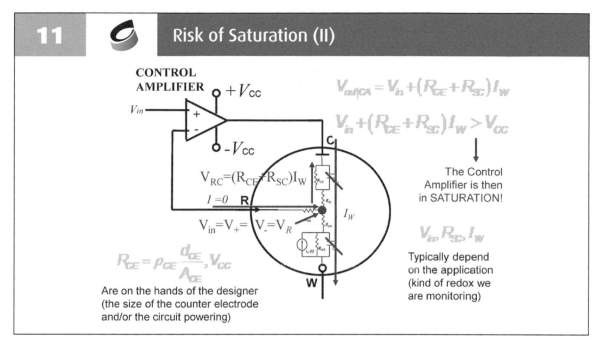

To avoid that, we need to design the electrochemical cell and, in particular, its CE larger enough to have the smallest possible R_{CE} to ovoid overcome the max available potential V_{cc}.

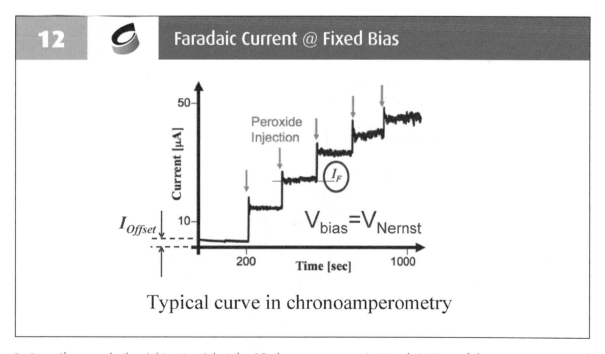

Typical curve in chronoamperometry

Now, if we apply the right potential at the RE, then we can acquire trends in time of the current generated at the WE (technique called *chronoamperometry*) and observe in such trends several current steps corresponding to any new injection of redox species in the chemical cell (e.g., the hydrogen peroxide). The current generated at the WE by redox reactions is called *Faradaic current*.

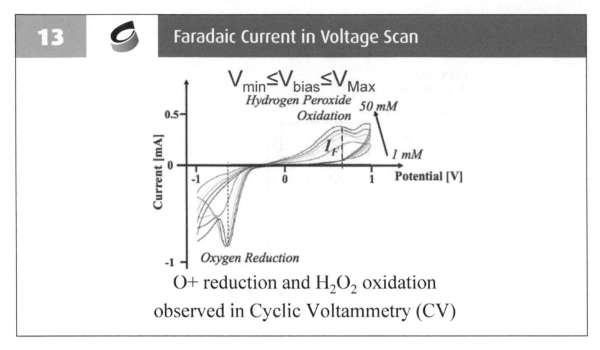

13 Faradaic Current in Voltage Scan

$$V_{min} \leq V_{bias} \leq V_{Max}$$

Hydrogen Peroxide Oxidation 50 mM

I_F

1 mM

Oxygen Reduction

O+ reduction and H_2O_2 oxidation observed in Cyclic Voltammetry (CV)

To know what is the right potential to be applied to the RE, we can acquire cyclic voltammograms, which typically present oxidation and reduction peaks at different potentials, called Nernst potentials. The Nernst potential is "the right potential" we need to apply at the RE to support the related redox reaction.

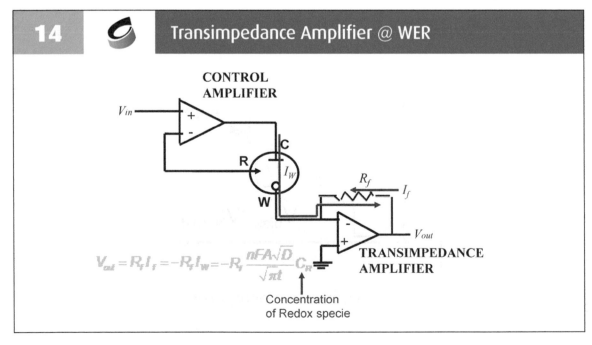

14 Transimpedance Amplifier @ WER

CONTROL AMPLIFIER

V_{in}

R_f I_f

I_W

W V_{out}

TRANSIMPEDANCE AMPLIFIER

$$V_{out} = R_f I_f = -R_f I_W = -R_f \frac{nFA\sqrt{D}}{\sqrt{\pi t}} C_R$$

Concentration of Redox specie

The acquisition of the Faradaic current may be performed by a current-to-voltage conversion with a transimpedance amplifier connected at the WE. In this way, the WE is virtually connected to ground (through the negligible potential between the "-" and "+" input of the OpAmp) while the output voltage is now proportional to the current flowing through the WE.

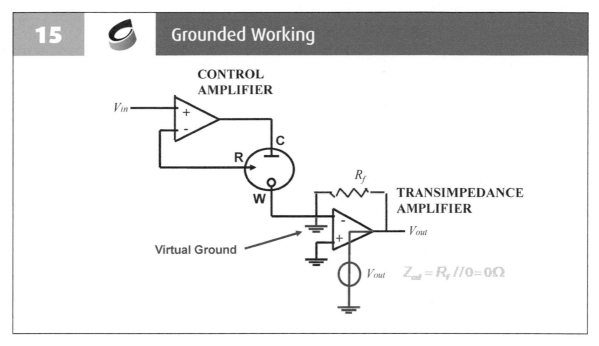

Another advantage of such a configuration is that the output impedance is null and, therefore, the reading of the electrochemical cell is done in a good manner with respect of the next block in the architecture of the whole acquiring system.

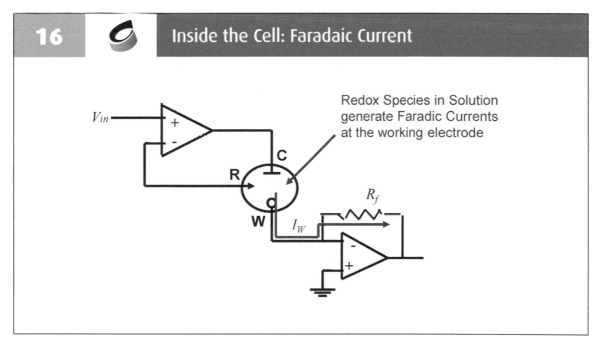

Of course, we need now to find a way to connect the redox reactions occurring in the electrochemical cell with the molecules that are markers of the diseases we would like to monitor in order to design the right electronic biosensors.

17 Outline

1. **Specificity by probe-proteins**

2. **Sensitivity by Carbon Nanotubes**

3. **Reliability in T & pH**

4. **Remote Powering**

5. **Minimally invasive**

To design the right electronic biosensors we need to provide right specificity and enough sensitivity to the biosensors, meanwhile assuring reliability with respect to tissues conditions (e.g., variations in temperature or pH) as well as minimally invasive solutions, which also require remote powering in many cases.

18 The challenges were...

1.**Specificity by probe-proteins**

2.Sensitivity by Carbon Nanotubes

3.Reliability in T & pH

4.Remote Powering

5.Minimally invasive

We will see in this chapter all these challenges, starting first with the several ways to provide the right specificity to the electrochemical sensors toward sensing aims in health monitoring.

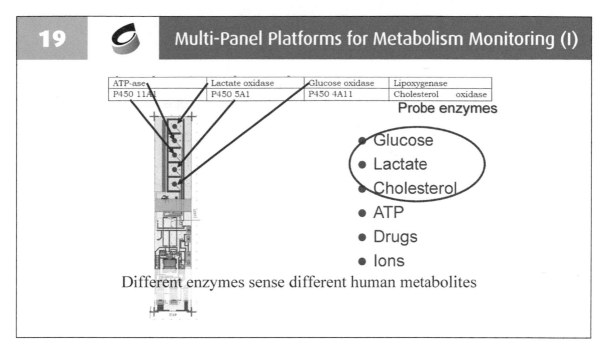

ATP-ase	Lactate oxidase	Glucose oxidase	Lipoxygenase
P450 11A1	P450 5A1	P450 4A11	Cholesterol oxidase

Probe enzymes

- Glucose
- Lactate
- Cholesterol
- ATP
- Drugs
- Ions

Different enzymes sense different human metabolites

There are many different molecules related to different diseases and, among them, a non exhaustive list includes glucose, lactate, cholesterol, Adenosine-TriPhosphate (ATP), several ions, and therapeutic drugs if we want to also develop new monitoring tools for personalised or precision medicine.

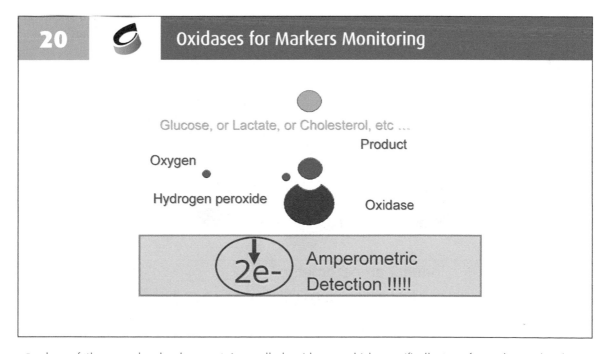

Glucose, or Lactate, or Cholesterol, etc ...

Oxygen

Product

Hydrogen peroxide

Oxidase

2e-   Amperometric Detection !!!!!

A class of these molecules has proteins, called oxidases, which specifically transform the molecules in metabolic pathways meanwhile producing hydrogen peroxide. At the WE, the hydrogen peroxide splits in hydrogen and oxygen by providing two electrons. Measuring these electrons provides an amperometric detection of the metabolite, the molecule, that has been transformed by the oxidase.

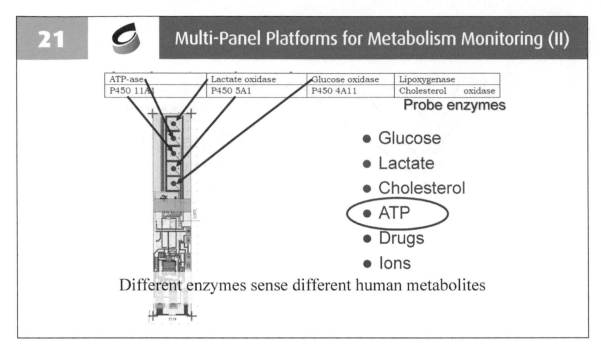

| ATP-ase | Lactate oxidase | Glucose oxidase | Lipoxygenase |
| P450 11A1 | P450 5A1 | P450 4A11 | Cholesterol oxidase |

21 — Multi-Panel Platforms for Metabolism Monitoring (II)

Probe enzymes

- Glucose
- Lactate
- Cholesterol
- ATP
- Drugs
- Ions

Different enzymes sense different human metabolites

Another class of molecules, and among them the ATP, does not have oxidases but has another class of proteins that help in defining an amperometric method for their detection.

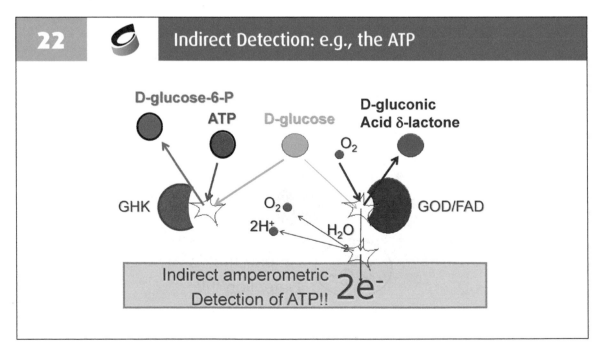

22 — Indirect Detection: e.g., the ATP

D-glucose-6-P ATP D-glucose D-gluconic Acid δ-lactone

O₂

GHK O₂ GOD/FAD

2H⁺ H₂O

Indirect amperometric Detection of ATP!! 2e⁻

Let us consider, as an example, the case of ATP: it has the Glucose HexoKinase (GHK) that transform the ATP by consuming glucose too. However, the transformation based on HGK does not provide any electron, while the co-presence of the glucose oxidase at the same interface provides hydrogen peroxide and, therefore, the possibility of an indirect detection of the ATP.

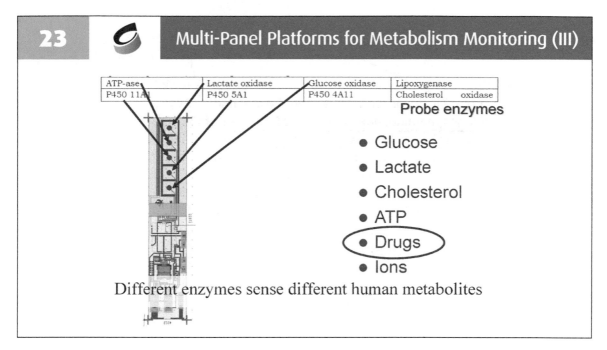

| ATP-ase | Lactate oxidase | Glucose oxidase | Lipoxygenase |
| P450 11A1 | P450 5A1 | P450 4A11 | Cholesterol oxidase |

Probe enzymes

- Glucose
- Lactate
- Cholesterol
- ATP
- Drugs
- Ions

Different enzymes sense different human metabolites

A third class of molecules, exogenous metabolites, allows amperometric detection: the one of therapeutic compounds. In this case, we can as well design a specific detection of therapeutic compounds toward the development of technologies for personalised or precise medicine.

P450 for Drugs Monitoring

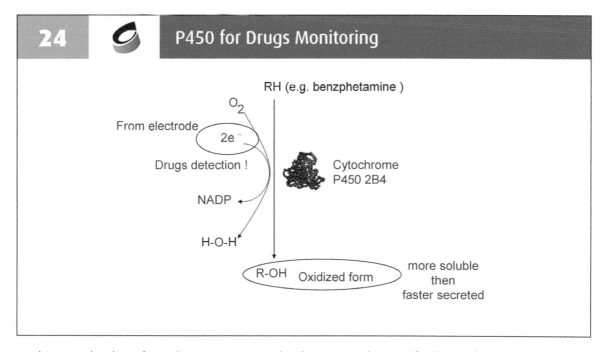

In this case, the class of cytochromes P450 provides the proteins that specifically transform the target drugs. For example, the benzphetamine is transformed by the isoform 2B4 of the cytochromes P450, which oxidase the drug by consuming two electrons. Once again, providing these two electrons from the WE and measuring the related current provides an amperometric method for a specific measure of the drug.

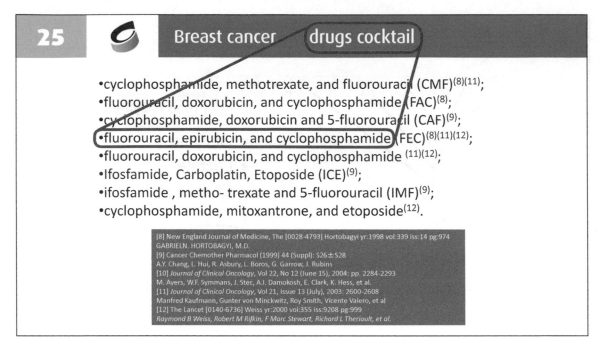

However, the typical problem in detecting drugs is that the therapies are usually based on series of different drugs provided simultaneously while the cytochromes P450 are not specific as well as than oxidases. Therefore, there is the need to develop special approaches to improve the specificity at system level.

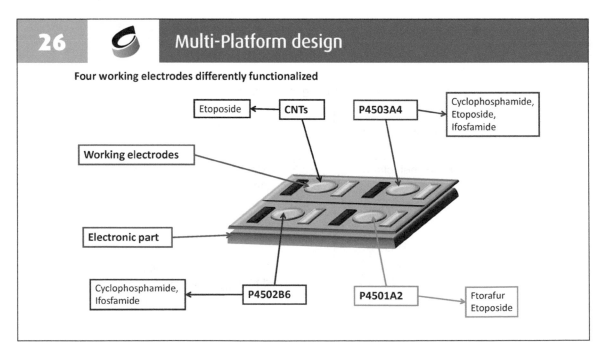

For example, if we design a platform to detect several drugs based on three different isoforms of the cytochromes P450 to detect four different drugs, then we can query the different amperometric biosensors in a way to successfully cross-correlate their measures to improve specificity.

27 Multiple Calibration Curves

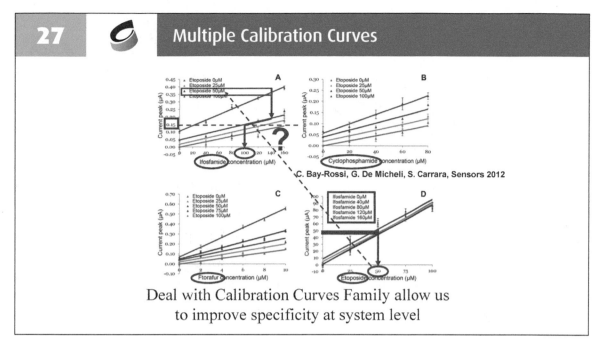

C. Bay-Rossi, G. De Micheli, S. Carrara, Sensors 2012

Deal with Calibration Curves Family allow us
to improve specificity at system level

In fact, the measure of one drug (e.g., the Ifosfamide) is cross-correlated with the measure of Etoposide. Likely, the Etoposide is not correlated with the measure of Ifosfamide and, therefore, a first measure on the Etoposide enables to calibrate the measure on the Ifosfamide and, thus, to correctly estimate the concentration of both Etoposide and Ifosfamide.

28 Sensors Query in Time

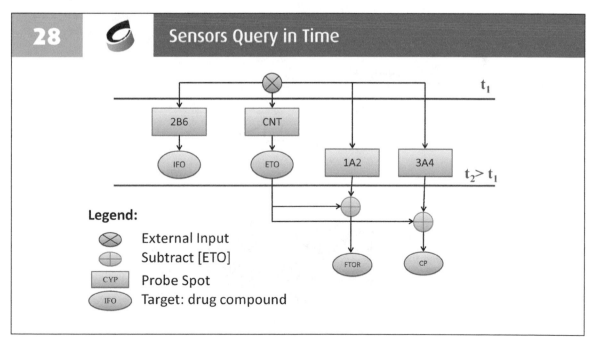

So, as a general approach, we can develop systems that have different queries to sensors in time, in order to acquire, first, those sensing information that are useful, in a successive instant of time, to correctly calibrate the measures on a further series of measure on other sensors of the same system.

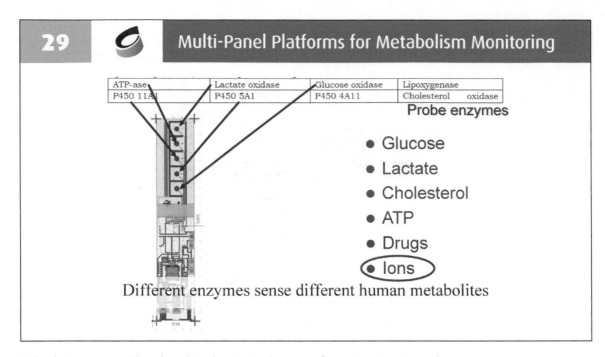

The last case considered in this chapter is the one of sensing ions. In such a case, we cannot use an amperometric method to acquire data about ions, while we can perform potentiometric measures.

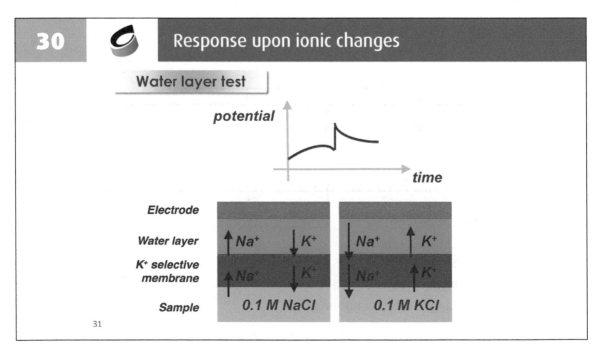

The potentiometric measure is a common procedure to detect the formation of an aqueous layer between the selective membrane and the electrode upon ionic changes in solution. For example, when a K+ selective membrane is exposed to a solution of Na+, ion-exchange processes at the interface cause changes in ion composition and a measure of the interface potential registers such a change.

31 The challenges were...

1. Specificity by probe-proteins

2. **Sensitivity by Carbon Nanotubes**

3. Reliability in T & pH

4. Remote Powering

5. Minimally invasive

Once we have identified the right method for specifically measure a biomarkers, we need then to assure the right sensitivity to the biosensors in order to cope with the typical concentration ranges foreseen by the diseases we would like to monitor. To that aim, several kind of nanomaterials are extremely useful to improve the sensing performance of the electrochemical interfaces.

32 Problems on Detection Limits

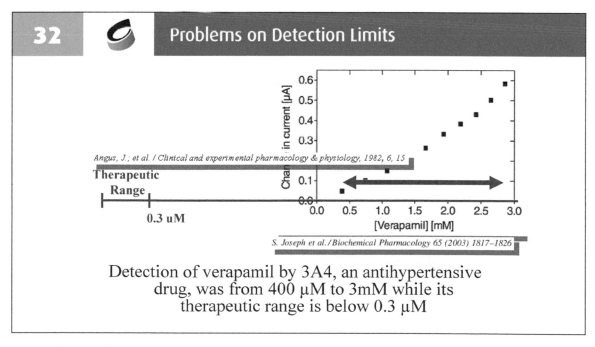

Angus, J.; et al. / Clinical and experimental pharmacology & physiology, 1982, 6, 15

Therapeutic Range

0.3 uM

S. Joseph et al. / Biochemical Pharmacology 65 (2003) 1817–1826

Detection of verapamil by 3A4, an antihypertensive
drug, was from 400 μM to 3mM while its
therapeutic range is below 0.3 μM

To see the issue here, we need to understand that is not sufficient to measure the target molecule in a good range (e.g., the Verapamil in the range from 0.5 to 3 mM), but we need to target the range usually find in patient' liquids (e.g., below 0.3.M for the Verapamil in human blood).

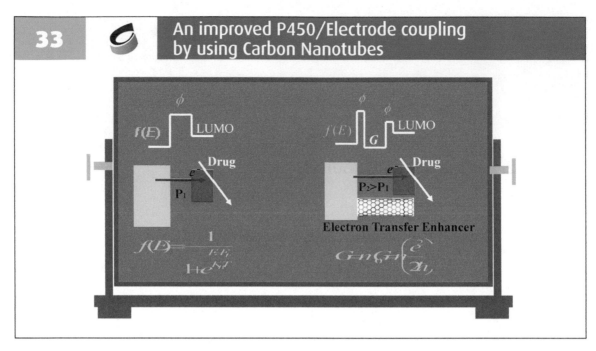

One of the reasons of such limited sensitivity is the tunnelling barrier between Lowest Un-Occupied Molecular Orbital (LUMO) in the redox molecule and the Fermi level in our electrodes. The presence of electron mediators (e.g., carbon nanotubes) can support the electron transfer and increase the rate of electrons exchange between the redox molecules and the electrode.

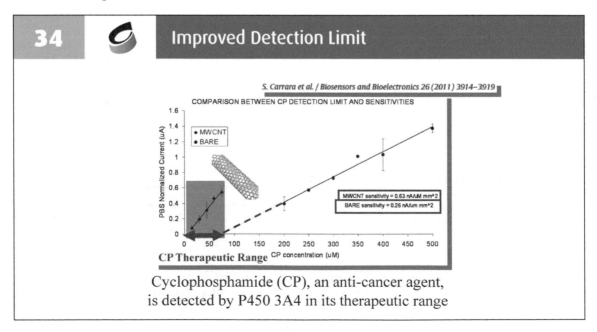

Cyclophosphamide (CP), an anti-cancer agent,
is detected by P450 3A4 in its therapeutic range

In some cases, it might be the difference between having or not a reliable technology for the considered application. For example, the cyclophosphamide (a very well-known anti-cancer compound) is detected with the cytochromes P450 3A4 even without Multi Walled Carbon Nano-Tubes (MWCNT) in the electrochemical interface but not in a suitable range, while the presence of the nanostructures assure a good fit with the therapeutic range of the drugs.

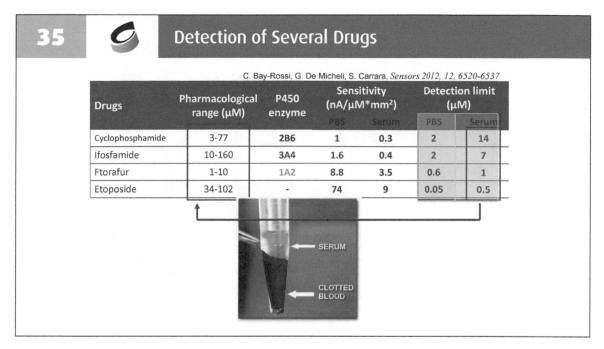

35 — Detection of Several Drugs

C. Bay-Rossi, G. De Micheli, S. Carrara, *Sensors 2012, 12, 6520-6537*

Drugs	Pharmacological range (µM)	P450 enzyme	Sensitivity (nA/µM*mm²)		Detection limit (µM)	
			PBS	Serum	PBS	Serum
Cyclophosphamide	3-77	2B6	1	0.3	2	14
Ifosfamide	10-160	3A4	1.6	0.4	2	7
Ftorafur	1-10	1A2	8.8	3.5	0.6	1
Etoposide	34-102	-	74	9	0.05	0.5

SERUM

CLOTTED BLOOD

Thank to the use of MWCNT it has been possible to successfully detect several anti-cancer therapeutic compounds (e.g., Cyclophosphamide, Ifosfamide, Ftorafur, and Etoposide) in the pharmacological ranges and in blood serum.

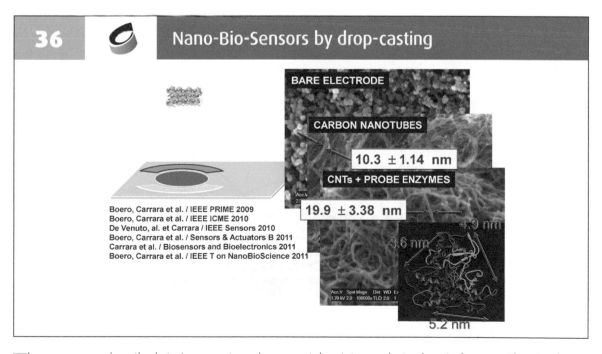

36 — Nano-Bio-Sensors by drop-casting

BARE ELECTRODE

CARBON NANOTUBES

10.3 ± 1.14 nm

CNTs + PROBE ENZYMES

19.9 ± 3.38 nm

4.9 nm

3.6 nm

5.2 nm

Boero, Carrara et al. / IEEE PRIME 2009
Boero, Carrara et al. / IEEE ICME 2010
De Venuto, al. et Carrara / IEEE Sensors 2010
Boero, Carrara et al. / Sensors & Actuators B 2011
Carrara et al. / Biosensors and Bioelectronics 2011
Boero, Carrara et al. / IEEE T on NanoBioScience 2011

There are several methods to incorporate carbon nanotubes into an electrochemical sensor. The simplest is the drop-casting, where the nanotubes previously dispersed typically in chloroform are just drop-casted onto the WE previously to deposit the specific sensing probe-protein. Of course, this method works very well in those cases where the WE is large enough for a direct hand-manipulation.

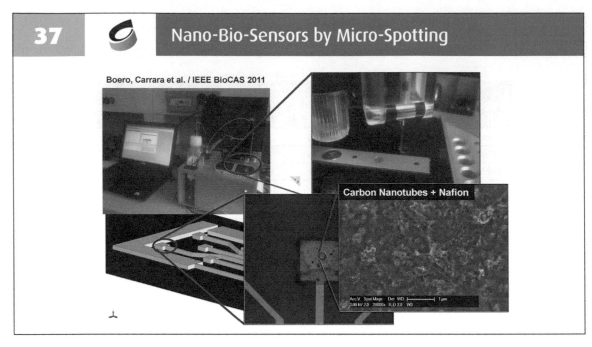

In those cases where, instead, the WE is in micron sizes, then a micro-spotter may be use for spotting the nanotubes precisely on top of the WE only. One drawback of such a method is the fact that micro-spotters usually spot water-based solutions and this force the dispersion of carbon nanotubes (usually hydrophobic) in water in combination with a water-soluble polymer (e.g., the nafion).

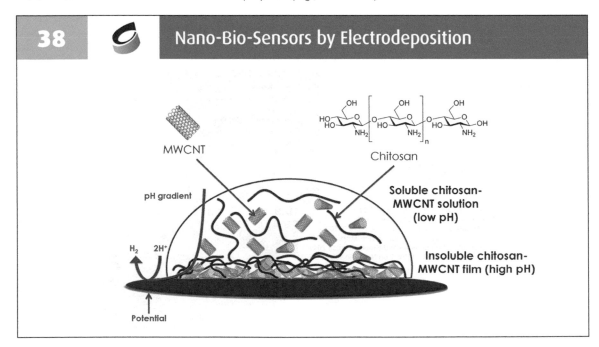

An alternative method is the electrodeposition. In this case, by applying a fixed potential to the electrode, the pH in proximity of the surface gradually change to higher values, and this change in pH triggers the chitosan polymerization that creates a 3D matrix able to entrap the MWCNT.

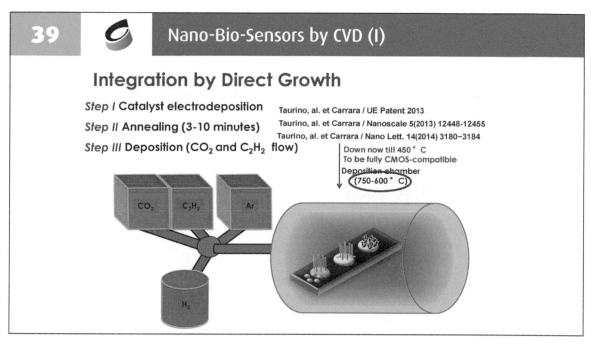

A further method is the direct grown by Chemical Vapor Deposition (CVD). Here, the sensors device is located in a deposition chamber heated up at the desired temperature. Then, after some minutes of annealing under hydrogen and Argon flow, carbon precursors were introduced in the deposition tube to fabricate the nanomaterials only on the WEs where the precursor film has been electrochemically previously formed.

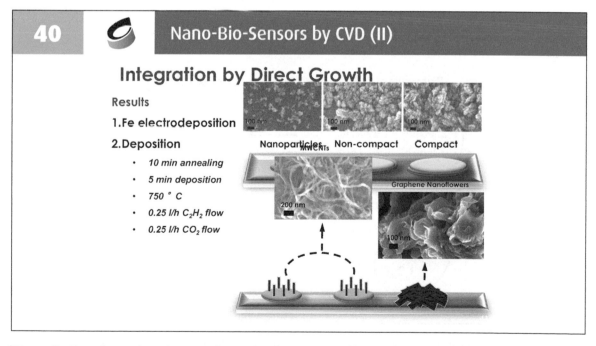

B y adjusting electro-deposition conditions for the precursor film (Fe-based), different kinds of precursor layers might be obtained (compact, non-compact, or in form of nano-particles) that then give different kind of carbon nanomaterials (long-and-tiny or short-and-large nanotubes, or graphite nanoflowers).

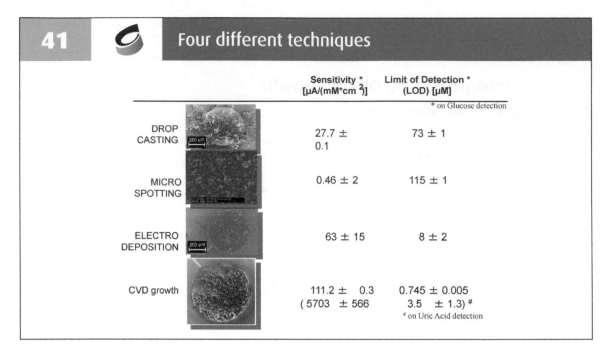

	Sensitivity * [μA/(mM*cm²)]	Limit of Detection * (LOD) [μM]
		* on Glucose detection
DROP CASTING	27.7 ± 0.1	73 ± 1
MICRO SPOTTING	0.46 ± 2	115 ± 1
ELECTRO DEPOSITION	63 ± 15	8 ± 2
CVD growth	111.2 ± 0.3 (5703 ± 566	0.745 ± 0.005 3.5 ± 1.3) # # on Uric Acid detection

The four presented techniques for carbon nanostructures deposition return different sensitivities and, therefore, different limits of detection (LOD). Among them, the CVD growth returns the highest sensitivity (then, the lowest LOD) since it assures the best electrical connection between the nanostructures and the underneath electrode surface.

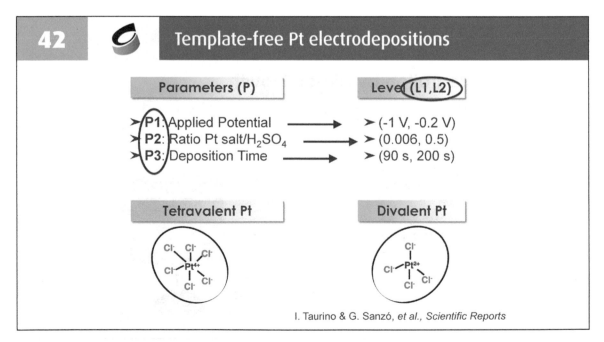

I. Taurino & G. Sanzó, *et al.*, *Scientific Reports*

Another kind of nanostructures we can use for improving the performance of our sensing systems is that based on nano-platinum as obtained by electrodeposition. The deposition can start with both tetravalent or divalent platinum salts in solution, which furnish the platinum atoms to be then clustered in nano-corals, nano-particles, or nano-flowers by adjusting the electrodeposition conditions (e.g., by changing the applied potential, deposition time, etc.).

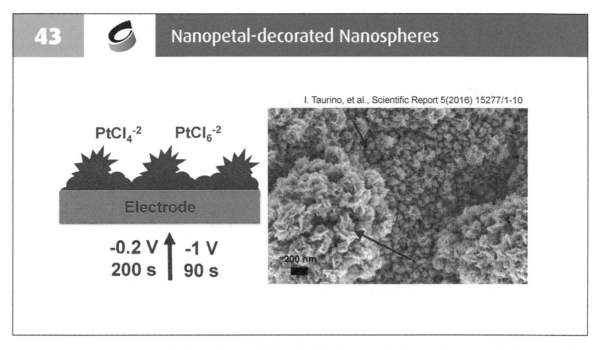

For example, 200 seconds of deposition at -200 mV with divalent platinum-salt returns Pt-Nanoparticles, while 90 seconds of deposition at -1 V with tetravalent platinum-salt returns Pt-nano-petals.

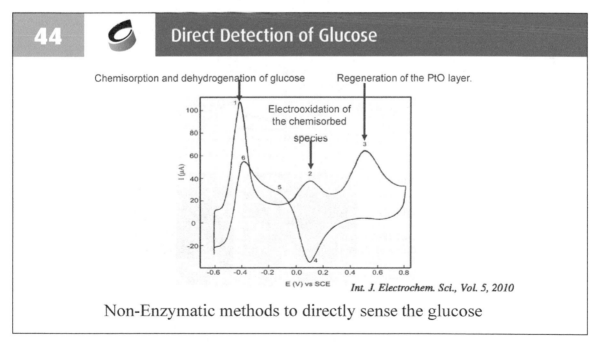

With this kind of nano-Pt is possible to also directly detect the glucose without any probe-protein (without any oxidase) since the glucose goes through a chemisorption and dehydrogenation at the nano-Pt surface that returns three oxidation peaks useful for the quantitative determination of the glucose concentration.

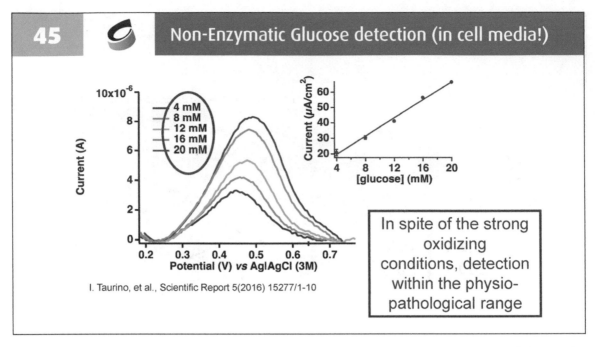

In fact, it is possible to detect the glucose in human blood as well as in cells-culture media by following the increase in current of one of these oxidation peaks as correlated with an increased amount of glucose in the samples.

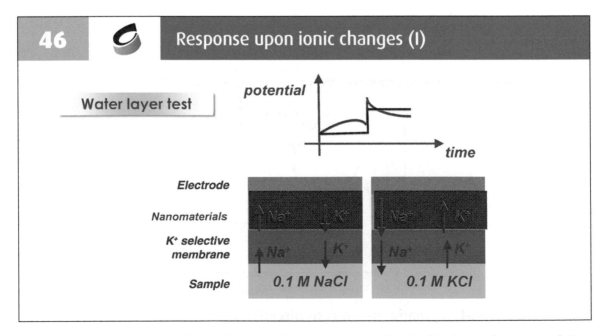

As already mentioned, potentiometric measure is a common procedure to detect ionic changes in solution. However, a presence of an non-avoidable small aqueous layer between the selective membrane and the electrode usually gives quite evident time trends in the measured potentials, which makes the technique not so suitable for real applications. Instead, this water layer is absent in presence of nanostructured Pt and, therefore, no drifts are observerd.

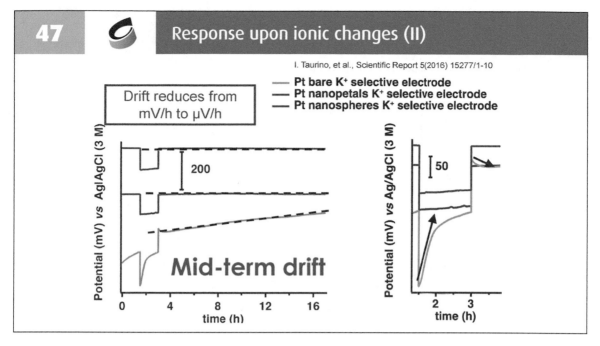

47 — Response upon ionic changes (II)

I. Taurino, et al., Scientific Report 5(2016) 15277/1-10

— Pt bare K⁺ selective electrode
— Pt nanopetals K⁺ selective electrode
— Pt nanospheres K⁺ selective electrode

Here the clear proof that the time-tend usually presents in potential measure acquired on bare Pt electrodes are totally absent (within a time window of more than 10 hours) thank to the presence of Pt nano-petals or nano-spheres onto the electrode surface.

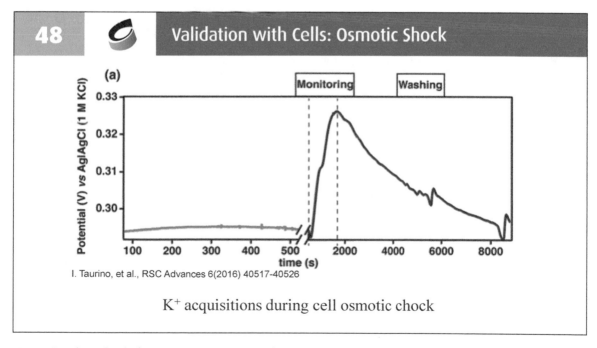

48 — Validation with Cells: Osmotic Shock

I. Taurino, et al., RSC Advances 6(2016) 40517-40526

K⁺ acquisitions during cell osmotic chock

By using these kind of nano-structures-powered potassium sensors it has been possible to reliably measure the potassium released in the culture medium by a cell culture during an osmotic shock.

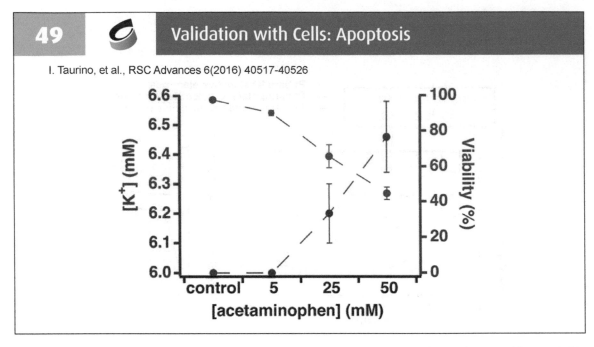

49 **Validation with Cells: Apoptosis**

I. Taurino, et al., RSC Advances 6(2016) 40517-40526

By using these kind of nano-structures-powered potassium sensors it has been also possible to reliably measure the potassium released in the culture medium by a cell culture during an intoxication due to injections of acetaminophen in toxic doses.

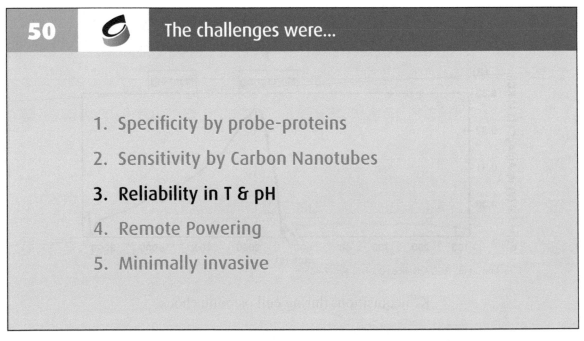

50 **The challenges were...**

1. Specificity by probe-proteins

2. Sensitivity by Carbon Nanotubes

3. **Reliability in T & pH**

4. Remote Powering

5. Minimally invasive

In order to successfully develop monitoring systems for human health, we need to assure reliable detection also dealing with respect variations of the hosting body dealing with temperature and pH variations.

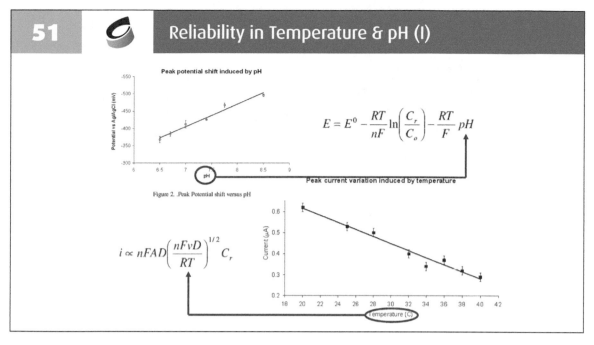

The reason is that both Nernst (for potentials) and Randle-Sev.ic (for currents) equations lead the behaviour of our electrochemical sensors and show dependence by temperature and/or pH.

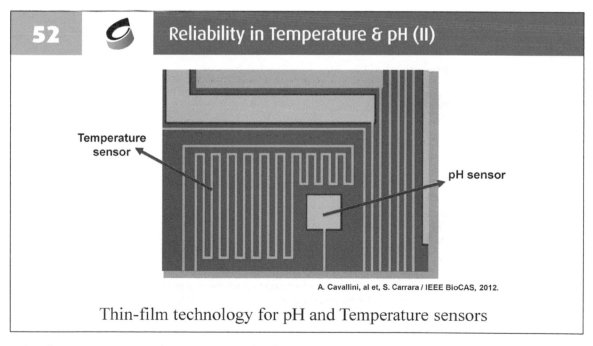

Therefore, temperature and pH sensors need to be integrated in our sensing interfaces in order to add the acquisition of these two ambient parameters for further calibrations on data then acquired by the specific electrochemical biosensors. So, for example, a temperature sensor and a pH meter may be realised at the electrochemical interface by using thin-films technologies by thermal evaporation of platinum.

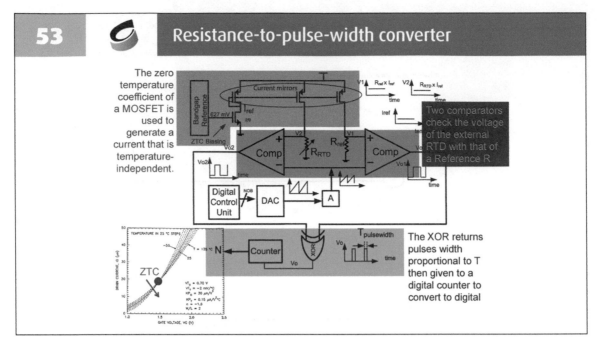

Both of these sensors need specific electronic readers. For example, a precise measure of temperature may be realised by comparing the voltage drop across the external Resistance Temperature-Dependant device (RTD) exposed at the interface with an internal reference resistor, both driven by a reference current and read by comparators with sawtooth thresholds to generate pulses with width proportional to the temperature.

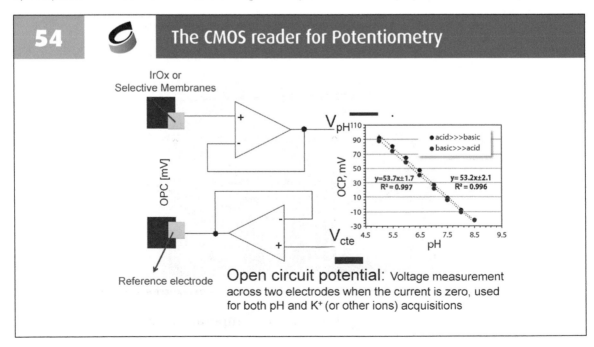

For the pH, as well as for monitoring ions, the electronic interface may be realised by using a voltage follower to read the Open Circuit Potential (OPC) at an platinum electrode with Iridium Oxide (IrOx, for pH) or with a selective membrane (for ions) in order to acquire the potential with reference to the RE already present at the interface for the other electrochemical biosensors.

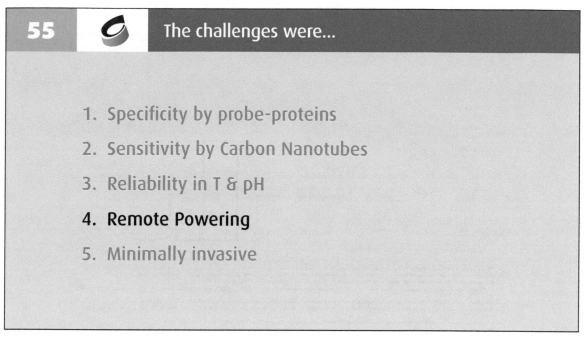

Once all the sensors at the interface are designed, the next step will be the design of the system required for the remote powering of the whole monitoring device.

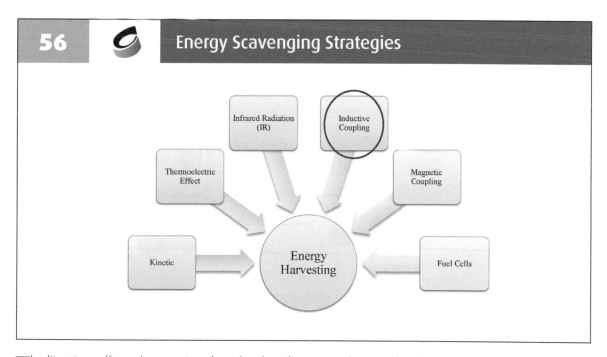

The literature offers a huge series of articles describing several approaches for powering electronic systems, including but not limited to kinetics, thermal, infrared radiation, inductive coupling, magnetic coupling, fuel cells.

57 — Inductive Coupling

Ref.	Coil Area (a = 10 mm²)	Carrier Frequency	Data Transmission	Bit Rate	Power Consumption	Efficiency	Distance	Measurement Site	Implantation Site
[8]	Tx: 7.8 λ Rx: 1.7 λ	4 MHz	twd Int.: PWM-ASK twd Ext.: ASK	twd Ext.:125 kbps	10 mW		5 mm	Air	Neural Recording System
[9]	Tx: 196.3 λ Rx: 31.4 λ	4 MHz	twd Ext.: LSK	5 kbps	6 mW		25 mm	Water Bearing Colloids	Various
[10]	Tx: 13200 λ Rx: 25.2 λ	1 MHz			150 mW	1% (min)	205 mm	PVC Barrel	Stomach
[11]	Tx: 184.9 λ Rx: 10 λ	1 MHz			10 mW	18.9% (max)	5 mm	Air	Cerebral Cortex
[12]	Tx: 282.7 λ Rx: 31.4 λ	0.7 MHz	twd Int.: ASK twd Ext.: LSK	twd Int.: 60 kbps twd Ext.: 60 kbps	50 mW	36% (max)	30 mm		Orthopaedic Implant
[13]	Tx: 31.4 λ Rx: 5 λ	10 MHz	twd Int.: ASK twd Ext.: BPSK	twd Int.: 120 kbps twd Ext.: 234 kbps	22.5 mW in vitro ≈ 19 mW in vivo		15 mm	Rabbit	Muscles
[14]	Tx: 196.3 λ Rx: 3.5 λ	5 MHz	twd Int.: OOK	100 kbps	5 mW		40 mm		Neural Stimulator
[15]	≈ Rx: 112.5 λ	6.78 MHz	twd Int.: OOK twd Ext.: LSK	twd Ext.:200 kbps	120 mW	20% (max)	25 mm	Dog Shoulder	Muscolar Stimulator
[18]	Tx: 40 λ Rx: 0.4 λ	915 MHz			0.14 mW	0.06%	15 mm	Bovine Muscle	Various

[8] T.Akin et al., "A wireless implantable multichannel digital neural recording system for a micromachined sieve electrode", IEEE J. Solid-State Circ., vol.33, pp. 109-118, Jan 1998
[9] C.Sauer et al., "Power Harvesting and Telemetry in CMOS for Implanted Devices", IEEE Trans on Circuits and Systems, vol.52, n.12, pp.2605-2613, 2005
[10] B. Lemaerts et al., "An inductive power link for a wireless endoscope", Biosensors and Bioelectronics, vol.22, pp. 1890-1895, 2007
[11] K.M. Silay et al., "Load Optimization of an Inductive Power Link for Remote Powering of Biomedical Implants", IEEE Proc. of International Symposium on Circuits and Systems 2009, pp. 533-536, May 2009.
[12] B. Lemaerts et al., "An inductive power system with integrated bi-directional data-transmission", Sensors and Actuators A, vol. 115, pp.221-229, 2004
[13] J. Parramon et al., "ASIC-based battery less implantable telemetry microsystem for recording purposes", Eng. in Med. and Bio. Soc., In Proc. of the 19th Annual Int. Conf., vol.5, pp. 2225-2228, 1997.
[14] G. Gudnason et al., "A Chip for an implantable Neural Stimulator", Analog Integrated Circuits and Signal Processing, vol.22, pp.81-89, 1999
[15] B. Smith et al., "An externally powered, multichannel, implantable stimulator-telemeter for control of paralyzed muscle", IEEE Trans. on Biomed. Eng., vol.45, pp.463-475, 1998
[18] A.S.Y.Poon et al., "A mm-sized implantable Power Receiver with Adaptative Link Compensation", Stanford University

Among these several, inductive link looks the right one for the applications we are targeting here since it has been already successfully demonstrated be capable to supply power through distances of several tens of mm, both by simulations and experiments, with efficiencies around 20%.

58 — Measures on the Designed Inductors

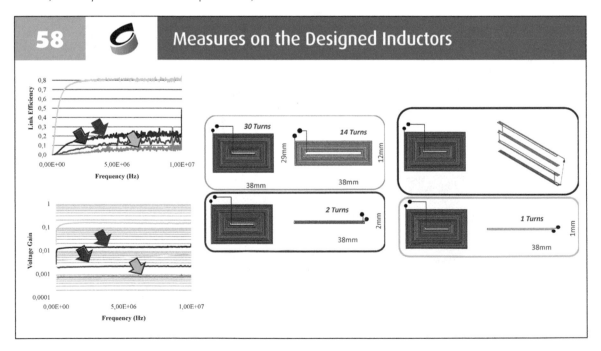

Of course, great challenge is the design of the receiving coil because smaller it is and smaller will be the received power. Trying minimizing the lateral size of that coil means dropping down the power efficiency. A good compromise may be obtained by designing receiving coils with several turns by exploiting the z-dimension (multi-layer coils).

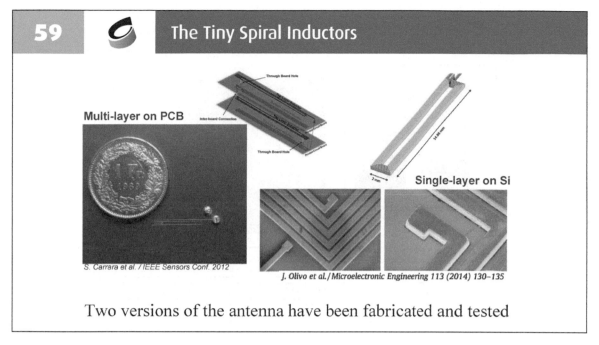

multi-layer coils exploiting the z-dimension may be realized on a multi-foil Printed Circuit Board (PCB). An alternative is represented by the possibility for micro-fabricated coils directly realized in a silicon substrate. In this second case, the power efficiency is augmented by realizing several turns, laterally organized, by in micro-sizes.

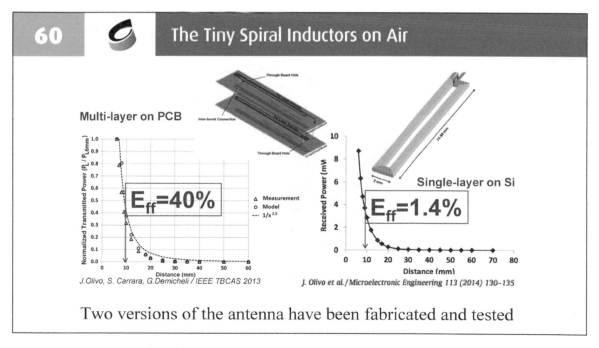

However, experiments have demonstrated much better capability to transfer power if the receiving coil is realised on multi-layer PCB with respect to coils micro-fabricated on silicon. In the first case, the power transfer efficiency has reached 40% in air.

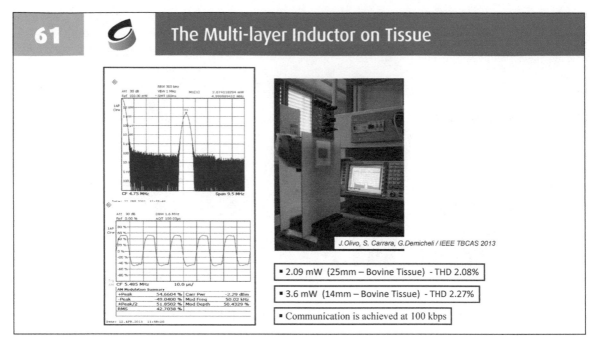

61 The Multi-layer Inductor on Tissue

J.Olivo, S. Carrara, G.Demicheli / IEEE TBCAS 2013

- 2.09 mW (25mm – Bovine Tissue) - THD 2.08%
- 3.6 mW (14mm – Bovine Tissue) - THD 2.27%
- Communication is achieved at 100 kbps

By through bovine tissues, the multi-layer coil realized on PCB has succeeded in receiving up to 2.09 mW through 25 mm and 3.6 mW through 14 mm of bovine tissue.

62 The Realized Remote Powering Patch (I)

J.Olivo, S. Carrara, G.De Micheli / IEEE TBCAS 2013

The patch has been realized with off-the-shelf components

On the other side of the inductive link, a multi-turn transmitting coil may be realized without particular geometrical limitation because it has to be located on the patch foreseen on top of the skin. This patch may host enough large powering coil and still has space for all the required components for the powering and data-transmission electronics.

63 — The Realized Remote Powering Patch (II)

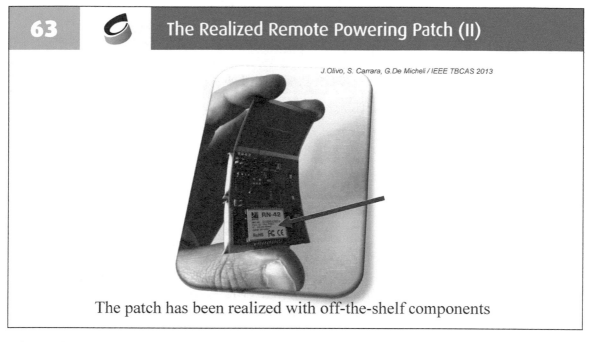

J.Olivo, S. Carrara, G.De Micheli / IEEE TBCAS 2013

The patch has been realized with off-the-shelf components

The patch on top of the skin has space enough for also host a Bluetooth communication system for data transmission toward a mobile phone or a tablet.

64 — The Android Interface (I)

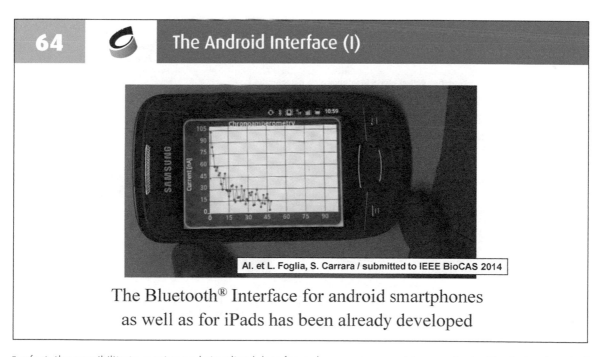

Al. et L. Foglia, S. Carrara / submitted to IEEE BioCAS 2014

The Bluetooth® Interface for android smartphones
as well as for iPads has been already developed

In fact, the possibility to receive and visualized data from chronoamperometries acquired with electrochemical biosensors has been demonstrated with Android™ smart phones....

65 — The Android Interface (II)

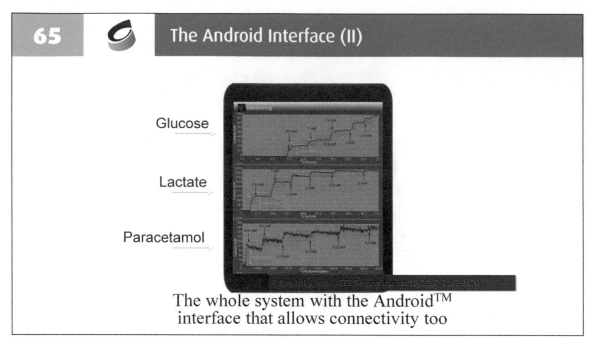

The whole system with the Android™
interface that allows connectivity too

• • • .as well as with Android™ tablets. So, the monitoring of endogenous (e.g., glucose and lactate) and exogenous (e.g., paracetamol) metabolites have been demonstrated with Android™ portable devices.

66 — Connectivity with Smart-Watch

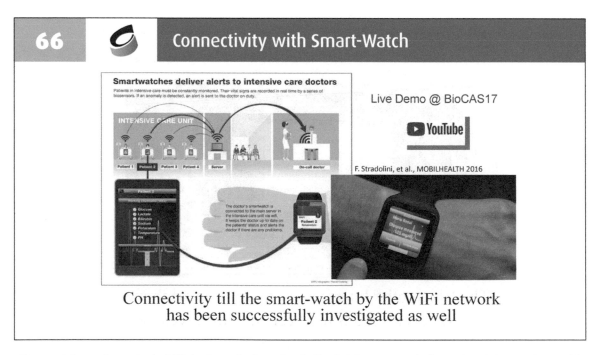

Connectivity till the smart-watch by the WiFi network
has been successfully investigated as well

Once data are in an Android™ device, a further step is the link to a connected smart-watch, e.g., to provide alerts and warnings to the professional personnel of intensive care units.

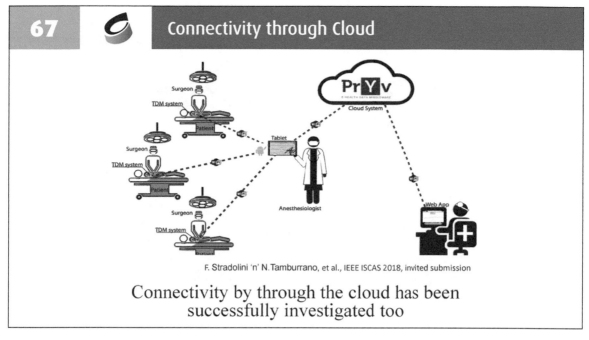

67 — Connectivity through Cloud

F. Stradolini 'n' N. Tamburrano, et al., IEEE ISCAS 2018, invited submission

Connectivity by through the cloud has been
successfully investigated too

On the other hand, once data are in an Android™ device, a further step is the connection to cloud in order to enable professional personnel of hospitals and private cabinets to remotely access all the patients' data.

68 — The challenges were...

1. Specificity by probe-proteins

2. Sensitivity by Carbon Nanotubes

3. Reliability in T & pH

4. Remote Powering

5. **Minimally invasive**

Last step of the design is now the CMOS frontend, which has to fulfil the final requirement of size that are minimally invasive. In this respect, the extremely high scale of integration (e.g., sub 100 nm) is not necessarily required since the "old CMOS technologies" still present plenty of room to host all the required electronics for our aims.

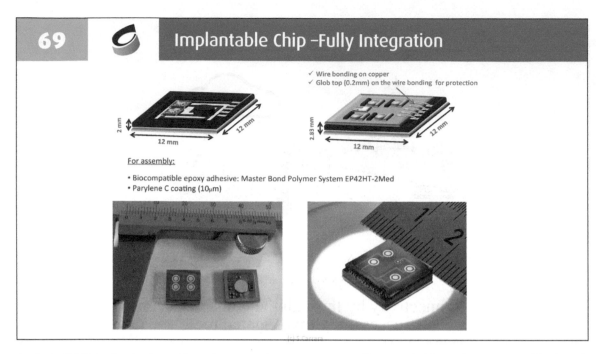

In fact, CMOS technologies with nodes at 0.35 or 0.18.m are still very reliable to design CMOS block with quite limited sizes (let say in few mm²) and that have space for the whole frontend for all the sensors in the designed platform, both in the case of implanted devices with global sizes of around 1 cm², and....

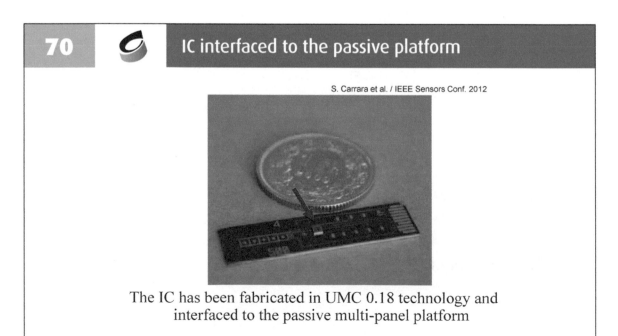

... also in the case of smaller implantable devices with sizes and shapes more similar to ones of a surgery needle.

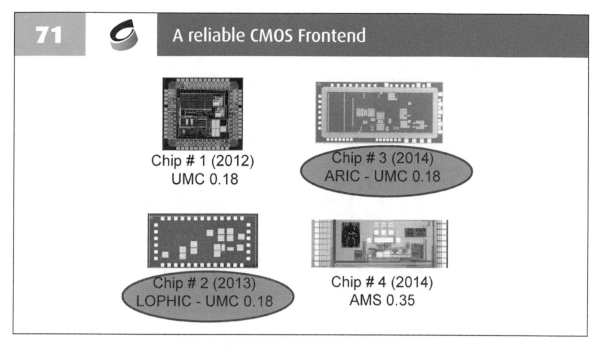

71   A reliable CMOS Frontend

Chip # 1 (2012)
UMC 0.18

Chip # 3 (2014)
ARIC - UMC 0.18

Chip # 2 (2013)
LOPHIC - UMC 0.18

Chip # 4 (2014)
AMS 0.35

To demonstrate the possibilities still offered by such CMOS technologies, we have realized, over the years, several prototypes of the frontend electronics.

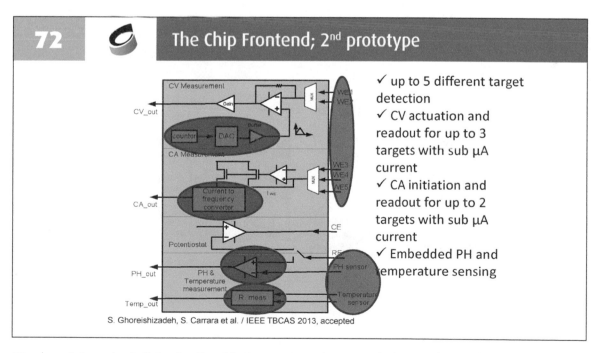

72   The Chip Frontend; 2nd prototype

✓ up to 5 different target detection
✓ CV actuation and readout for up to 3 targets with sub μA current
✓ CA initiation and readout for up to 2 targets with sub μA current
✓ Embedded PH and temperature sensing

S. Ghoreishizadeh, S. Carrara et al. / IEEE TBCAS 2013, accepted

Such prototypes host all the functional features we need to correctly drive and read all the sensors of the implantable platforms, including voltage followers for reading the open circuit potentials, amplifiers to read currents from working electrodes, signal generation to provide both fixed and cyclic reference voltages.

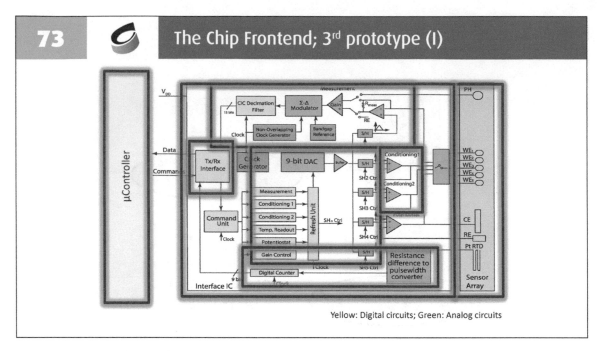

73 — The Chip Frontend; 3ʳᵈ prototype (I)

Yellow: Digital circuits; Green: Analog circuits

In some cases, the design of the prototype also implements a sigma-delta modulator to provide the Analog-to-Digital converter before sending data to the microcontroller by through the I/O interface. The cyclic voltages generated to support cyclic voltammetry are obtained with a mixed-signal design to implement Direct Digital Synthesis Methods.

74 — The Chip Frontend; 3ʳᵈ prototype (II)

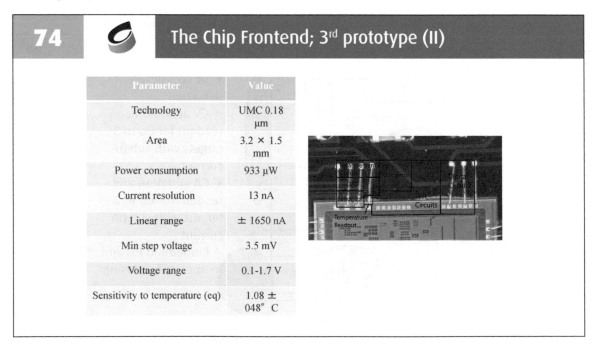

Parameter	Value
Technology	UMC 0.18 μm
Area	3.2 × 1.5 mm
Power consumption	933 μW
Current resolution	13 nA
Linear range	± 1650 nA
Min step voltage	3.5 mV
Voltage range	0.1-1.7 V
Sensitivity to temperature (eq)	1.08 ± 048° C

As an example, we can see here the characteristics of one of the Integrated Circuits (ICs) fabricated in UMC 0.18.m CMOS technology, which consumes 900 μW (with 1.8V supply voltage). This power consumption is well below the upper limits of 3.6 mW reached with the realized inductive link and, therefore, the possibility to power such kind of minimally invasive frontend circuits is then demonstrated.

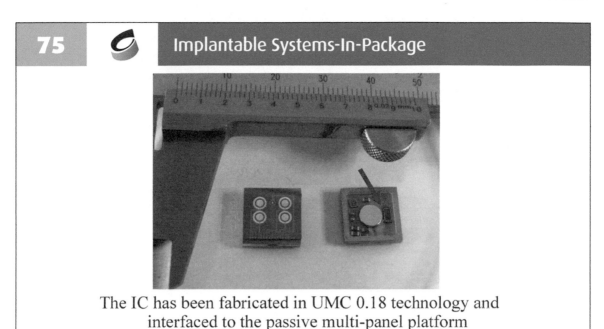

Implantable Systems-In-Package

75

The IC has been fabricated in UMC 0.18 technology and interfaced to the passive multi-panel platform

Here on the right (indicated by the read arrow) the IC integrated in one of the implantable device, while on the left side is the top layer with in evidence the four independent electrochemical sensors and, less visible, the pH and the temperature sensors in the middle.

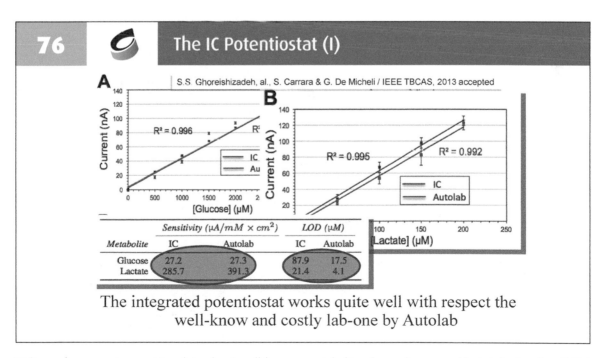

The IC Potentiostat (I)

76

S.S. Ghoreishizadeh, al., S. Carrara & G. De Micheli / IEEE TBCAS, 2013 accepted

	Sensitivity ($\mu A/mM \times cm^2$)		LOD (μM)	
Metabolite	IC	Autolab	IC	Autolab
Glucose	27.2	27.3	87.9	17.5
Lactate	285.7	391.3	21.4	4.1

The integrated potentiostat works quite well with respect the well-know and costly lab-one by Autolab

The performance in acquiring data about well-known metabolites (e.g., glucose and lactate) of such a CMOS IC are quite good with respect the acquisition done with a top lab-equipment (a potentiostat from Autolab). The IC returns very similar sensitivities, while worst LODs are due to a high noise level in data acquired with the designed IC.

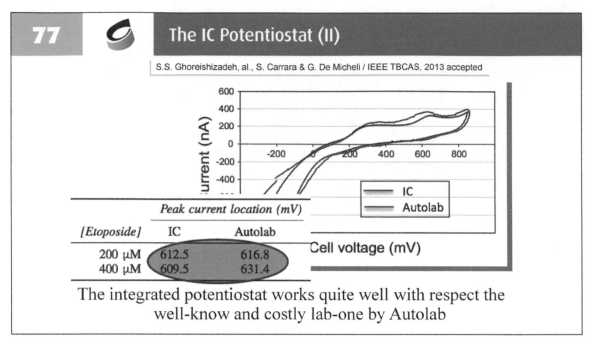

77 The IC Potentiostat (II)

S.S. Ghoreishizadeh, al., S. Carrara & G. De Micheli / IEEE TBCAS, 2013 accepted

[Etoposide]	Peak current location (mV)	
	IC	Autolab
200 μM	612.5	616.8
400 μM	609.5	631.4

The integrated potentiostat works quite well with respect the well-know and costly lab-one by Autolab

The performance are good as well in sensing exogenous therapeutic compounds (e.g., the Etoposide), where the right location of the faradaic peaks are of key importance to also improve the specificity. Here the comparison with top lab-equipment (again an Autolab instrument) show very similar peak current location.

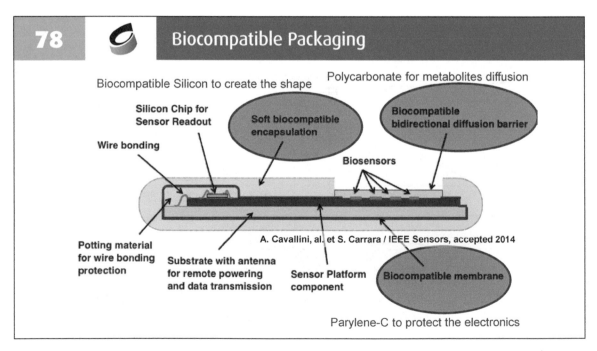

78 Biocompatible Packaging

A. Cavallini, al. et S. Carrara / IEEE Sensors, accepted 2014

A further step in the design is the transfer of biocompatibility before implantation. This may be provided by several kind of membranes, each one providing a different function: the Parylene-C is going to protect the electrical circuits from the host-body' liquids, the Polycarbonate allows the penetration of metabolites for their measure, and the biocompatible silicon assures the biocompatibility to the full packaging.

The biocompatible silicon cannot cover the whole device because it does not provide permeable membranes. Therefore, a open window is left in the silicon packaging just over the biosensors in order to let a further transfer of a polycarbonate membrane. An alternative is possible by using an epoxy-resin, which may be used to fully cover the whole device since it provides a permeable membrane. However, the biocompatibility of epoxy-resin is less than polycarbonate.

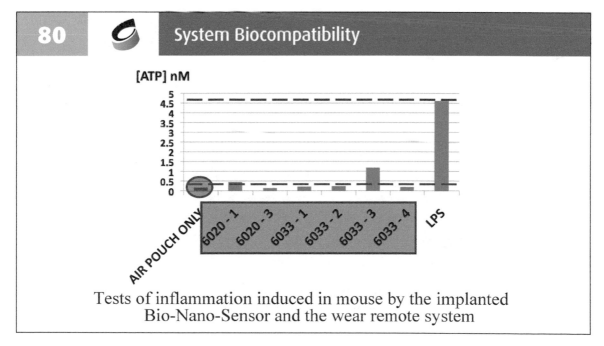

By using the biocompatible silicone, one month of implantation in mice returned a negligible level of inflammation (e.g., as monitored by measuring ATP) with respect to an inflammation artificially induced in the region of the implant (e.g., induced by injecting Lipopolysaccharides, LPS).

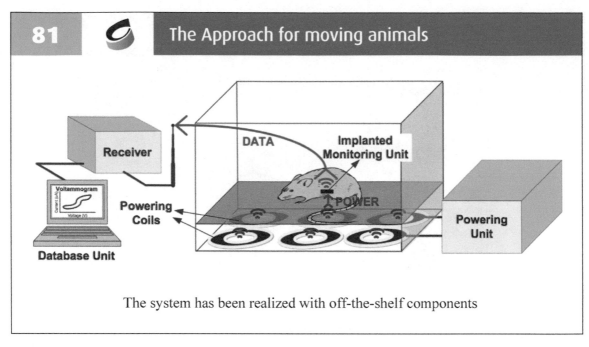

81 The Approach for moving animals

The system has been realized with off-the-shelf components

This kind of implantable device has been thought to be tested and validate with implantation in mice with possibility of remote powering from the cage bottom. Two strategies have been proposed: powering the implanted device with a series of powering coils equally located in the cage' floor, or...

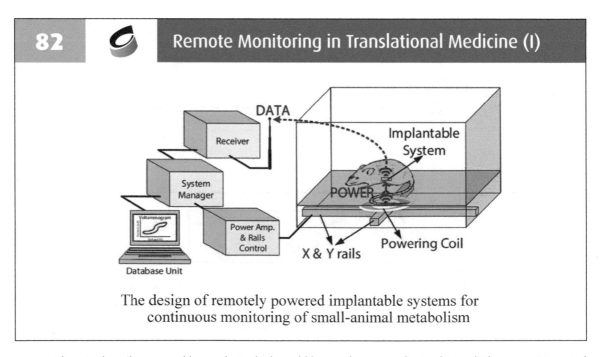

82 Remote Monitoring in Translational Medicine (I)

The design of remotely powered implantable systems for
continuous monitoring of small-animal metabolism

... with a single coil managed by a robot which could locate the mouse by tracking a little magnet inserted in the implantable device.

83 — Remote Monitoring in Translational Medicine (II)

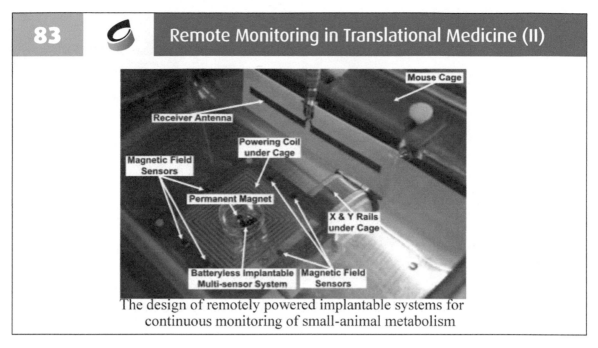

The design of remotely powered implantable systems for continuous monitoring of small-animal metabolism

In vitro experiments have been conducted to test all the functions of both the sensing implant as well as of the remote powering systems.

84 — Under the skin system

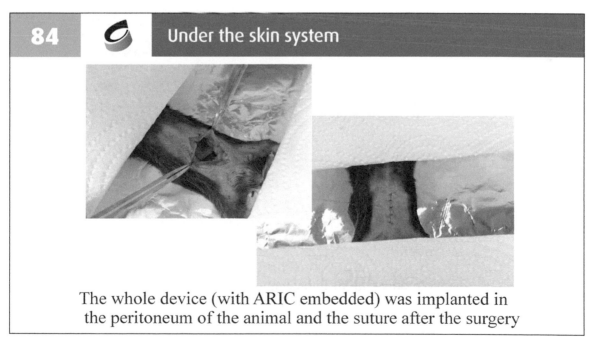

The whole device (with ARIC embedded) was implanted in the peritoneum of the animal and the suture after the surgery

And then a series of prototypes have been implanted in several mice for an in-vivo validation of the whole acquisition chain, from acquisition with biosensors to data visualisation on a screen.

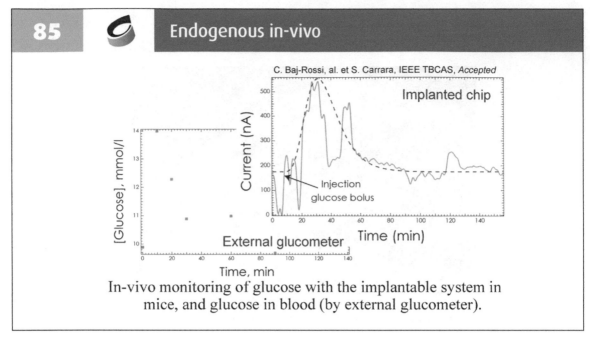

85 **Endogenous in-vivo**

C. Baj-Rossi, al. et S. Carrara, IEEE TBCAS, *Accepted*

Implanted chip

Injection glucose bolus

External glucometer

In-vivo monitoring of glucose with the implantable system in mice, and glucose in blood (by external glucometer).

The system validation was successful and data have been acquired on endogenous metabolites (e.g., glucose) as well as acquisition compared with measure done on the blood sampled form the mouse' tail and then glucose values obtained with a commercially-available glucometer.

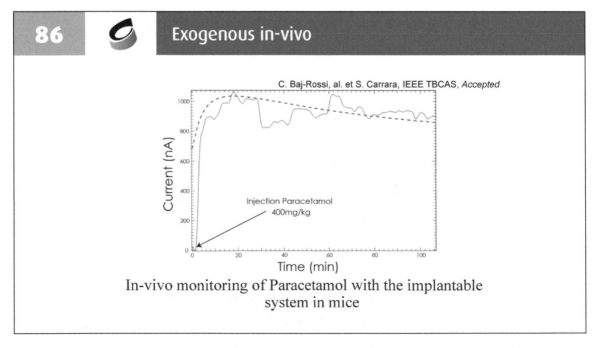

86 **Exogenous in-vivo**

C. Baj-Rossi, al. et S. Carrara, IEEE TBCAS, *Accepted*

Injection Paracetamol
400mg/kg

In-vivo monitoring of Paracetamol with the implantable system in mice

The system validation was successful also on de acquisition of data about exogenous metabolites (e.g., paracetamol) where typical pharmacologic curves have been acquired.

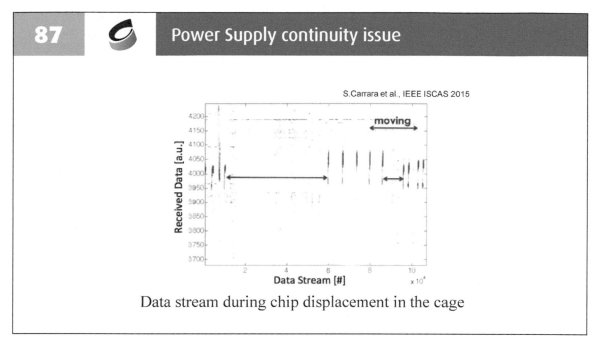

S.Carrara et al., IEEE ISCAS 2015

Data stream during chip displacement in the cage

Both in the case of endogenous and exogenous metabolites, the experiments for the validation have highlight a problem in the continuity of the power supply: when the mouse move too fast and the robot lost tracking it, then the powering system needs several seconds to find again the mouse, meanwhile loosing data streams.

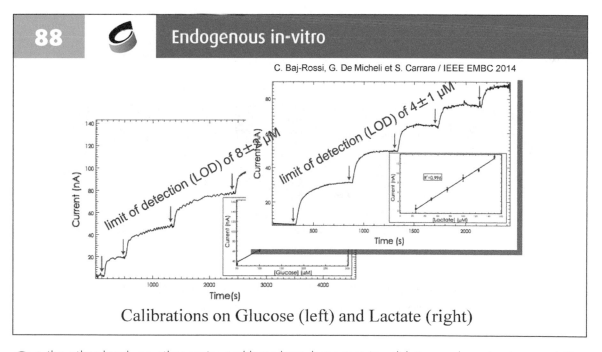

C. Baj-Rossi, G. De Micheli et S. Carrara / IEEE EMBC 2014

Calibrations on Glucose (left) and Lactate (right)

On the other hand, no other major problems have been met in validating such an in-vivo acquisition system. In fact, in-vitro validations (where the position of the powering coil with respect the receiving one has been strictly controlled) have always shown highly reliable acquisitions on endogenous metabolites (e.g., glucose and lactate) as well as than ...

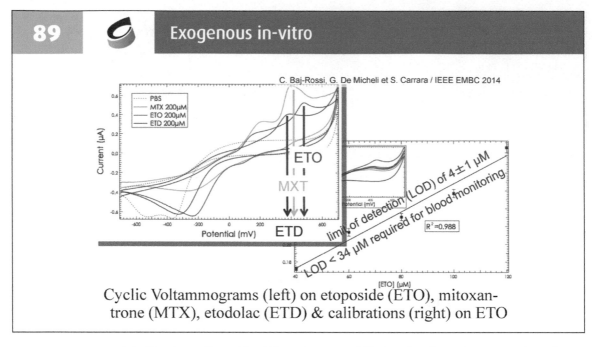

89 Exogenous in-vitro

Cyclic Voltammograms (left) on etoposide (ETO), mitoxantrone (MTX), etodolac (ETD) & calibrations (right) on ETO

• • • exogenous metabolites (e.g., Etoposide, Mitoxantrone, and Etodolac), where the system has always shown LODs well below the minimum concentrations registered in medical therapies.

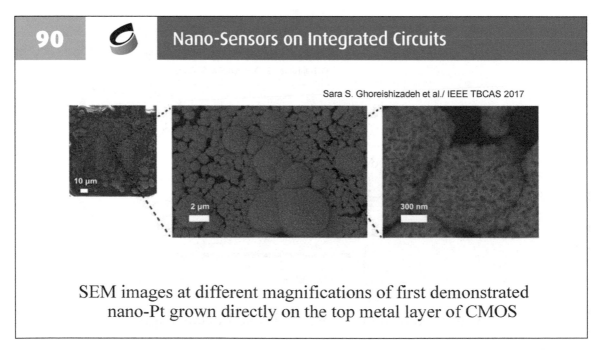

90 Nano-Sensors on Integrated Circuits

SEM images at different magnifications of first demonstrated nano-Pt grown directly on the top metal layer of CMOS

Last design step we could envisage in the development of highly innovative implantable and remotely-powered minimally-invasive devices is the integration of nanostructure direct on the top and exposed metal layer of our ICs. In this case, nano-Pt have been demonstrated a suitable material for direct nano-structuration on ICs as it is shown here by Scanning Electron Imaging.

91 A certain attention from international media

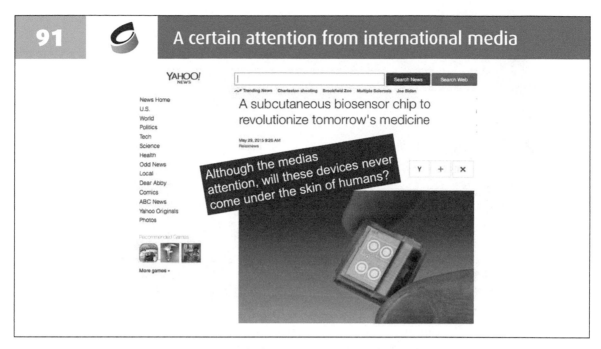

Of course, our successful developments, scientific findings, and bright validations obtained over the last years have also attracted a certain attention from international media. However, the question now is if these kind of implantable devices will never become medical tools usually implanted under the skin of humans.

92 Under the skin for body sculpting

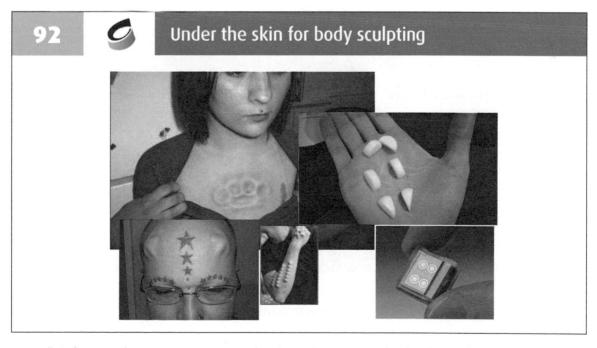

Well, in fact, new human generations are already used to accept under-the-skin implant also only for body sculpting and beauty aims.

And, already in 2015, Sandro Portner (a Swiss young man) has asked to have implanted under-the-skin by a siring injection a couple of passive RFID devices in order to be capable to be automatically recognized by his mobile phone or his home' door (copy/past in a browser the following link:
http://www.euronews.com/2015/06/23/rfid-chips-a-key-to-more-or-less-freedom).

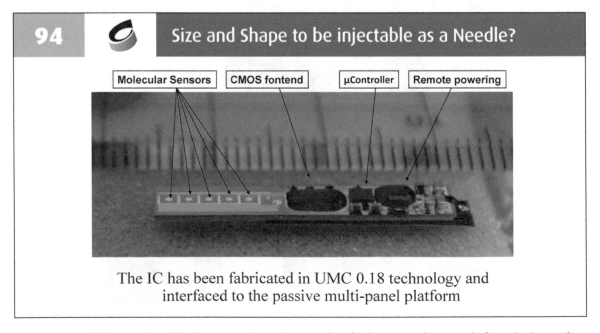

The IC has been fabricated in UMC 0.18 technology and interfaced to the passive multi-panel platform

The second of the implantable devices we have designed so far has sizes that exactly fit with those of an injectable surgery needle.

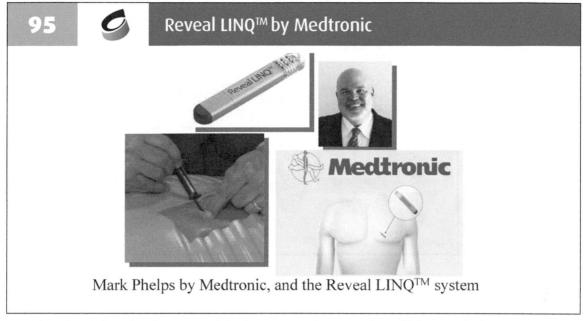

95   **Reveal LINQ™ by Medtronic**

Mark Phelps by Medtronic, and the Reveal LINQ™ system

And Medtronic has already presented the last version of its Reveal LINQ™ device, which is going to enter the market with an implantable procedure that exactly foreseen the injection by a siring. This is not an RFID passive device but instead a very active device to acquire ECG (Electrocardiogram) signals. Not yet a biosensing of molecular biomarkers but, for sure, very close.

96   **Conclusions**

- **CMOS design is required for a reliable Bio/CMOS Interfaces for Metabolism and, especially, electrochemical sensing under the skin**
- **Special Biotech solutions are necessary to target the right selectivity upon the target metabolic molecules**
- **Special Nanotech solutions are necessary to target the right sensitivity and limit-of-detection in the right ranges of metabolic concentrations**
- **Automatic and continuous monitoring of the metabolism in humans is actually feasible from body tissues to our portable electronics including smart-watches**

In conclusion: CMOS dedicated design is still required to realize reliable Bio/CMOS Interfaces for Metabolism monitoring with under-the-skin devices, special biotech solutions are necessary to provide the right selectivity while special nanotech solutions are necessary to assure the right sensitivity. With this approach, automatic and continuous monitoring of the metabolism in humans is actually feasible from body tissues to our portable electronics including smart-watches.

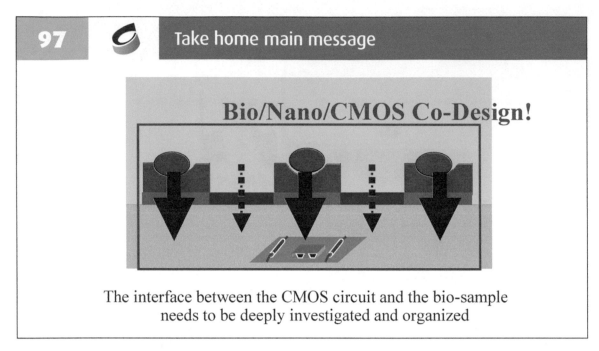

97 Take home main message

Bio/Nano/CMOS Co-Design!

The interface between the CMOS circuit and the bio-sample
needs to be deeply investigated and organized

Here the take-home-message is that we need to use this new co-design approach by taking into account the simultaneous design of the electronic frontend, the nano, and the bio layers of the system to successfully realize implantable and remotely-powered devices for applications in biosensing.

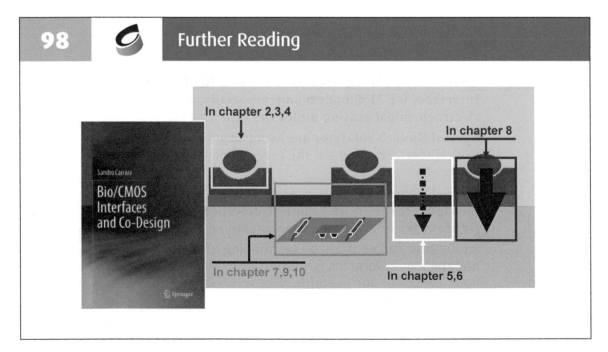

98 Further Reading

Sandro Carrara

Bio/CMOS
Interfaces
and Co-Design

Springer

In chapter 2,3,4

In chapter 8

In chapter 7,9,10

In chapter 5,6

For studying in details this new co-design approach it worth reading the book titled "Bio/CMOS interfaces and Co-Design" I have published in 2013, which has chapters addressing specific interactions of biomolecules to design different kind of biosensors, several nano-structured layers to improve the biosensors performance, and introduce several CMOS frontend architectures to design at best devices for medical applications

99   **Great thank to my team @ EPFL**

Of course, I have a great debt of gratitude with respect to my research group in EPFL as well as with the many friends and colleagues with which I have developed so many new bio/CMOS interfaces along last fifteen years of my carriers, ...

... and I thank you all for have followed this tutorial chapter about my research and the related new design approach. For any further information, please, do not hesitate to contact me or to have a look at my current research to follow further developments in the field.

CHAPTER 02

Optimization of the Transfer of Power and of the Data Communication in the Case of Remotely Powered Sensor Networks

Catherine Dehollain

Ecole Polytechnique Fédérale
de Lausanne (EPFL), Switzerland

My name is Catherine Dehollain, I am Professor in electronics at EPFL and I am responsible for the RFIC group. One of the main research topics of our group is dedicated to remote powering and data communication for sensor networks. We have started to work in this domain since year 2000 and we have got financial support thanks to the *Ecole Polytechnique Fédérale de Lausanne* (EPFL), to the Swiss National Research Foundation (SNF) for basic and fundamental research, to the Swiss CTI Office for applied research in collaboration with the Swiss industry as well as to the European commission for applied research in collaboration with other European universities, European research centers and European companies.

1		Content

- **PART 1: ARCHITECTURES OF REMOTELY POWERED SENSOR NETWORKS**

- **PART 2: PASSIVE TRANSMITTERS THANKS TO BACKSCATTERING DATA COMMUNICATION**

- **PART 3: REMOTE POWER FOR WIRELESS SENSOR NETWORKS**

Part 1 will be dedicated to the presentation of different architectures of remotely powered sensor networks.
Part 2 will be dedicated to explanations on passive transmitters for which the carrier signal, necessary to send the output electrical signals of the sensors through modulation of the carrier, is generated in the base station. This technique is of high interest to minimize the power consumption of the transmitter included in the sensor node.

Part 3 will be dedicated on explanations on remote powering operations by using either magnetic coupling (also called near field) or electro-magnetic coupling (also called far field).

PART 1

2		ARCHITECTURES OF REMOTELY POWERED SENSOR NETWORKS

Part 1 is dedicated to the architectures of remotely powered sensor networks.
It will be shown in this section that the electronic designer of such architectures has to be familiar with different topics which are important at architecture level such as: (i) topologies of receivers and transmitters to transfer data and choice of the best operating frequency, (ii) solutions to send power remotely to the sensor nodes and choice of the best operating frequency, (iii) choice of the best CMOS technology by taking into account key constraints such as the operating frequencies and the cost of the manufacturing process, (iv) design at transistor level of the different electronic blocks, (v) targeted dimensions and weight of the sensor node in line with the targeted application (e.g. Implanted Medical Device: IMD).

3

There are different possible topologies of receivers and transmitters for wireless communication at short distance. The choice of the topology will depend on key specifications at system level such as the distance range, the data rate, the power consumption.

There are different solutions to send power remotely to the sensor node. The choice of the solution will depend on key specifications at system level such as the targeted application (e.g. implanted medical devices, wireless sensor nodes on the body, environmental wireless sensor nodes, alarm systems, automotive sensor nodes, tire pressure monitoring systems: TPMS), the distance range, the temperature range, the supply voltage and the power consumption of the sensor node, the dimensions and the weight of the sensor node.

At the Boundary between Different Domains

- **New architectures of sensor nodes for wireless communications at short distance**
  - Back-scattering/ Load modulation (e.g. RFIDs), Impulse Radio Ultra Wideband (IR UWB), Super-regenerative transceivers, Direct conversion transceivers.
  - Biomedical field: implants, wearable sensor networks.
- **Remotely powered wireless circuits**
  - Through RF wave by magnetic coupling , electro-magnetic coupling, electro-acoustic coupling (ultrasound).
  - Rechargeable micro-batteries.
- **Ultra low-power wireless communications**
  - Low supply voltage imposed by advanced technologies, ultra-low current operation
- **MHz to GHz-range ICs**
  - RF models of the transistors and passive elements, parasitic coupling.
- **Fully integrated solutions**
  - RF and Mixed-mode circuits.
  - Circuits in advanced CMOS technologies.
  - Integrated tuning elements, MEMS (Micro Electro Mechanical Systems).

4

The first method uses one single frequency for remote power and data communication. The architecture of the transmitter of the sensor node is called "passive transmitter" as the RF carrier is generated in the base station. Therefore, it allows minimizing the power consumption of the transmitter. The basic principle is called respectively load modulation, for magnetic coupling, or backscattering modulation for electro-magnetic coupling between the sensor node and the base station. A trade-off has to be determined between power transfer and data communication because the same carrier signal is used for both operations.

Data Transfer Methods

- **One single frequency for remote power and data communication**
  - Passive transmitter in the sensor node (load modulation or backscattering)
  - Interruption of the transfer of power during data transmission
  - Optimisation of the link for power transfer and data communication thanks to trade-off

- **One frequency for data communication and another frequency for remote power**
  - Power transfer is independent of data communication
  - Extra power is required due to the active transmitter of the sensor node
  - Extra antenna is used for data communication

The second method uses one frequency for data communication and another one for remote power. By this way, the power transfer is independent of the data communication. The transmitter of the sensor node is called "active transmitter" because the RF carrier is now generated in the sensor node at the cost of a larger power consumption than the one of a passive transmitter.

Backscattering Modulation in far field

This method of modulation has been invented by Harry Stockman in year 1948.

The backscattering modulation works in far-field (electro-magnetic coupling). One single frequency is used for remote power and data communication.

It is now widely used for the identification of objects in the domain of Radio Frequency Identification Devices (RFID). The object contains a tag (also called transponder) which is tracked by the base station (also called interrogator or reader) if the tag is at a reasonable distance so that it can be remotely powered by the base station. The maximum distance

□ One single frequency for remote power and data communication

□ **Harry Stockman**, "Communication by Means of Reflected Power", **1948**

□ The tag is tracked if the base station (reader or interrogator) is in **range**

□ Minimization of the **power consumption** of the tag → Generation of the **carrier** at base station and **backscattering** of the incident wave

U. Karthaus and M. Fischer, "Fully integrated passive UHF RFID transponder IC with 16.7 uW minimum RF input power," IEEE Journal of Solid-State Circuits, vol. 38, pp. 1602–1608, Oct. 2003.
J.P. Curty, M. Declercq, C. Dehollain, N. Joehl, "Design and optimization of passive UHF RFID systems", Springer, year 2007

range depends on the operating frequency, on the power sends by the base station and on the gains of the antennas of the base station and of the tag.

The power consumption of the transmitter of the tag is low because the carrier is generated in the base station.

Load Modulation in near field

Load and backscattering modulations are similar because the carrier is generated in the base station for both modulations. As it is the case for backscattering modulation, one single frequency is used for remote power and data communication in this figure (Fig. 5.1 of the cited PhD thesis). The

One single frequency for remote power and data communication

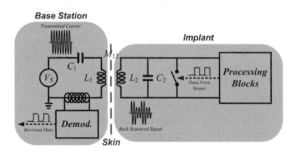

M. Azizighannad, "Remotely-Powered Batteryless Implantable Local Temperature Monitoring System for Freely Moving Mice", EPFL, PhD thesis no. 6253, Oct. 2014.

main difference between these two types of modulation is that load modulation works in near-field (magnetic coupling).

The maximum distance range between the base station and the implant depends on the operating frequency, on the power delivered by the base station and on the design of the primary and of the secondary coils.

Processing blocks correspond to the sensor, its electronic interface (low noise amplifier, A/D converter), the rectifier and the voltage regulator.

The data communication from the implant to the base station is called uplink data communication. The switch in parallel with the secondary coil is controlled through the data which correspond to the measurements of the sensor.

7   Wireless Active Transmitter

The transmitter works in far-field (electromagnetic coupling).

The carrier is generated in the tag thanks to the LC oscillator constituted by the two cross-coupled pairs of transistors, by the off-chip loop antenna and by the two capacitors C1 and C2 (Fig. 9 of the cited article). The loop antenna acts as the inductor of the LC tank and as an antenna. The two crossed-coupled pairs of transistors correspond to a negative conductor which is necessary to compensate the positive conductor (resistive loss) of the LC tank.

The transistor M1 at the bottom of the schematic is turned on and off thanks to the data signal. By this way, the On-Off Keying (OOK) amplitude modulation is implemented.

The Frequency Shift Keying (FSK) modulation could also be implemented by keeping M1 always on and by modifying the values of the two capacitors of the LC tank by the data signal. In such a case, the two capacitors are replaced by two banks of capacitors or by two varicaps.

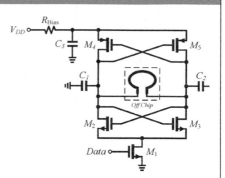

The transmitter works in far field.

The carrier is generated in the tag.

The frequency of the carrier is determined by the LC tank.

The **On-Off Keying (OOK)** amplitude modulation is used.

. Ghanad, M.M. Green, C. Dehollain, "A 30 uW Remotely Powered Local Temperature Monitoring Implantable Syst Transactions on Biomedical Circuits and Systems, vol. 11, no. 1, pp. 54-63, Feb. 2017.

8   Wireless Remote Powering

Wireless remote powering operates in magnetic coupling (near-field) when the distance between the sensor node and the base station is much smaller than the wavelength divided by 2 Pi. This wavelength depends on the operating frequency of the signal delivered by the base station which is dedicated to remote power. Frequencies of operation are dictated by the international radio regulations. The distance in air is of the order of 10 cm. Magnetic coupling between the primary coil of the base station and the secondary coil of the sensor node is highly sensitive to misalignment.

Wireless remote powering operates in electro-magnetic coupling (far-field) when the distance between the sensor node and the base station is much larger than the wavelength divided by 2 Pi. Frequencies of operation are dictated by the international radio regulations. The distance in air is of the order of 15 m. The data rate is normally higher than for magnetic coupling as the carrier frequency is much higher.

- ☐ **Magnetic Coupling (Near-Field)**
  - ☐ Near field region → d < λ/2π
  - ☐ Typical frequency bands: 125 kHz, 6.78 MHz, 13.56 MHz
  - ☐ Effective operation distance of 10 cm in air
  - ☐ Highly sensitive to misalignment between the primary coil and the secondary coil of the transformer
  - ☐ Higher energy efficiency in short distance at d<10 cm

- ☐ **Electro-magnetic Coupling (Far-Field)**
  - ☐ Far field region: → d > λ/2π
  - ☐ Typical frequency bands: 868 MHz, 915 MHz, 2.45 GHz, 5.8 GHz.
  - ☐ Effective operation distance up to 15 m in air
  - ☐ Higher data rate than the near-field systems

9   Single Frequency for Power and Data

The same frequency is used for remote power and data communication. In this figure (Fig. 3.1 of the cited PhD thesis), the assumption is done that the wireless system operates in far-field as two antennas are represented. This figure can be extended to near-field by replacing these two antennas by a trans-

O. Kazanc, "Far-Field Remotely Powered Wireless Sensor Platform for Real-Time Monitoring», EPFL, PhD thesis no. 6152, June 2014.

former. The reader is also called interrogator or base station. Downlink and uplink data communications respectively correspond to transfer of data from the reader to the sensor node and in opposite direction. The system is constituted by four blocks: (i) remote power block to send power from the reader to the sensor node, (ii) power management block to store the energy and to regulate the supply voltage, (iii) data acquisition block to measure (e.g.) the temperature and to digitalize this measurement through an A/D converter, (iv) transceiver block to exchange data between the sensor node and the reader.

10   Dual Frequency for Power and Data

System Building Blocks

→ Three fundamental units
1. Implantable sensor system
2. External base station for **data communication** and remote powering
3. Long distance **data communication** for data base and reporting

Two different frequencies are used for remote power and data communication. In this figure, the assumption is done that the blocks dedicated to remote power operate at 13.56 MHz in near field as two coupled coils are represented whereas the ones for data exchange work at 868 MHz in far field due to the two represented antennas.

The system is constituted by three units: (i) the sensor node which contains the remotely powered integrated circuit and the sensor, (ii) the two blocks of the base station for remote power and data communication, (iii) computers /mobile phones to store and to process the data.

The most important constraints are normally linked to the volume, to the dimensions, to the weight and to the power consumption of the sensor node.

The sensor can be for example an external thermistor to measure the temperature.

11

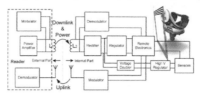

Knee Prosthesis Monitoring

Goals
- Increase of the life expectancy of the knee prosthesis
- Monitoring of the force, movement of the knee and temperature

Challenges
- Low coupling factor of inductive link due to limited secondary coil dimensions
- High power requirement (10 to 20 mW)

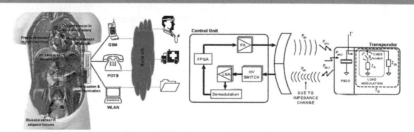

Swiss SNF NanoTera
Simos Project

O. Atasoy, C. Dehollain: "Remote Powering Realization for Smart Orthopedic Implants", 2012 NEWCAS Conference, pp. 521-524.
O. Atasoy, C. Dehollain, "Full-duplex Communication and Remote Powering Implementation of an Electronic Knee Implant", 2013 PRIME Conference, pp. 217-220.

The system block diagram is represented here (Fig. 1 of 2012 NEWCAS article). A frequency of 13.56 MHz is selected for remote powering. An inductive link power efficiency of 30% is achieved by using inductive powering at 2.5 cm. An overall remote powering efficiency of 14% is obtained to generate a supply voltage of 1.8 V for an output DC power of 20 mW. Another supply voltage of 2.8V is provided to the sensors by a CMOS switched-capacitor voltage doubler and a voltage regulator. The overall remote powering efficiency is equal to 12%.

A carrier frequency of 13.56 MHz is used for downlink data communication. The external power amplifier is designed as an Amplitude Shift Keying (ASK) modulator. The CMOS ASK demodulator of the implant can demodulate up to 1 Mbit/s of data-rate, while consuming 15 uW.

A carrier frequency of 432 MHz is chosen for the uplink wireless data communication. The Frequency-Shift-Keying (FSK) CMOS transmitter consumes 250 uW at a rate of 57 kbit/s.

12

Ultrasonic Powering and Data Communication

European FP7 project: www.ultrasponder.org

F. Mazzilli, C. Lafon, C. Dehollain, "A 10.5 cm Ultrasound Link for Deep Implanted Medical Devices", IEEE Transactions on Biomedical Circuits and Systems, vol. 8, no. 5, Oct. 2014, pp. 738–750.
F. Mazzilli, E.K. Kilinc, C. Dehollain, "3.2 mW Ultrasonic LSK Modulator for Uplink Communication in Deep Implanted Medical Devices", IEEE BIOCAS Conference, Oct. 2014, pp. 636-639.

The use of ultrasonic waves to power remotely an implanted medical device (IMD), and to exchange data with it, is very interesting in the case of deep implants (more than 5 cm inside the body) because the attenuation is much lower than for RF waves.

The frequency of the ultrasonic wave is chosen equal to 1 MHz for remote powering and bi-directional data communication.

For a transmitter-receiver distance of 10.5 cm, the overall power efficiency is characterized at 1 MHz in water using a phantom material. The transducers active area are equal to 3 cm* 9.6 cm (48 elements for the external control unit) and to 0.5 cm * 1 cm (1 element for the transponder). The power efficiency is equal to 1.6% by using a discrete components rectifier.

The uplink data communication is achieved by Load Shift Keying (LSK) modulation (Fig. 1 of the BIOCAS 2014 article). The total power consumption of the transponder (including the micro-controller) is equal to 3.2 mW in transmit mode and to 49 uW in sleep mode.

13

The patient swallows the pill to detect some problems in the gastrointestinal track. Its 3D position is detected thanks to the emission of a signal at 125 KHz through three class-D full bridge power amplifiers and three orthogonal coils included in the pill. The coils are powered on and off consecutively and in the same order. The emitted signals are detected by receivers and coils embedded in an intelligent jacket worn by the patient. Therefore, this system is like a GPS using magnetic coupling.

The diameter of the first version of the pill, which used discrete electronic components, was equal to 0.8 cm (Fig. 12 of the ISCAS 2015 article). It has

Digestive Track Diagnostic

→ Diagnosis of digestive system for:
• Constipation
• Irritable Bowel Syndrome (IBS)
• Gastroparesis

→ 3D trajectory information of the pill through the gastrointestinal track.

→ The pill provides three axis magnetic field for location information.

→ Fully integrated ASIC development enables miniaturization of the pill.

ASIC development

Ø 6mm   Ø 8mm

CTI Swiss Project

J. L. Merino and C. Dehollain, "LC tank full bridge control for large coil variations", 2012 ICECS conference, pp. 653-656.
J.L. Merino, O. Kazanc, N. Brunner, V. Schlageter, M. Demierre, C. Dehollain, "Low power receiver for magnetic digestive motility tracking pill", 2015 IEEE ISCAS Conference, pp. 459-472.

been reduced to 0.6 cm thanks to the design of the electronics in 0.18 um CMOS technology. This reduction enables the use of the tracking pill for pediatric patients.

Thanks to the CMOS implementation, the power consumption of the electronic pill has been reduced from 1.15 mW to 0.25 mW compared to the discrete electronic components design.

14

The reader sends power to the memory tag to power it remotely. The tag contains a memory to store information which is sent by the reader by downlink communication. Then, this information can be sent by uplink communication to another reader which is in the distance range of the tag.

A museum is a possible application. The director can update regularly the information linked to each object which is contained in the memory tag near the object. Then, the visitor can upload on his reader (e.g. part of his mobile phone) this information and use it afterwards at home.

The backscattering modulation is used for uplink

Passive Memory Tag for High Data Rate

Goal: to obtain a medium operating range between the reader and the passive memory tag for a given RF output power of the reader and a high data rate

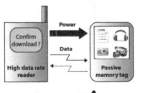

Read/write data rate: 10 Mbit/s
Distance range: 30 cm
Power consumption of the memory tag: 10 mW

Remote powering: 868 MHz (ISM)
Downlink data communication: 868 MHz (ISM)
Uplink data communication: 2.4 GHz (ISM)

MINAmi   **European IP Project MINAMI**

N. Pillin, N. Joehl, C. Dehollain and M. Declercq "High Data Rate RFID Tag / Reader Architecture using Wireless Voltage Regulation", IEEE TCAS1 Journal, vol. 57, no. 3, March 2010, pp. 714-724.

data communication. The same frequency than for remote power is used for downlink data communication.

Due to the regulations with respect to authorized frequency bands and levels of emitted power, the system specifications imply that a dual frequency architecture has to be selected: 868 MHz for remote power and downlink data communication, 2.4 GHz for uplink data communication.

15 Magnetically-Coupled Remote Powering System for Freely Moving Animals

This sub-section is dedicated to the implementation of a sensor node at a distance of 1 to 2 cm inside the body to measure (e.g.) some physiological parameters in the peritoneal region of a mouse.

Magnetic coupling between the external base station and the implant is selected to give power wirelessly to the implanted sensor node. The operating frequency is equal to 13.56 MHz. The fre-quency of the carrier is the same for downlink data communication.

An active CMOS transmitter is designed at 868 MHz for uplink data communication.

A low power time-based CMOS sensor interface is presented in the case of a thermistor to measure the temperature of the brown adipose tissue of a mouse.

16 Specs for Freely Moving Laboratory Rodents

Different types of biosensors are included in a biocompatible package which is implanted in the rodent. The implant is remotely powered by the external base station. The rodent is awake and it moves in a laboratory cage. The specifications with respect to a medium size mouse are considered.

- □ Application: Multi-bio sensor monitoring
- □ Condition of animal: mobile/awake
- □ Fully implantable: Yes
- □ Weight: Less than 2 g (10% of animal weight)
- □ Volume: Less than 1 cm³
- □ Maximum overall power consumption: 2 mW
- □ Powering type: Wireless Power Transfer (WPT)
- □ Remote powering distance: 3 cm
- □ Uplink data communication rate: 100 kbit/s
- □ Data communication distance: 40 cm (worst case)

The targeted power consumption of the implant is equal to 2 mW.

The weight and the volume of the implant should be respectively lower than 10% of the weight of the mouse and lower than 1 cm3.

The distance ranges between the external base station and the implant are respectively equal to 3 cm for remote power, and to 40 cm for uplink data communication.

The data rate for uplink communication is equal to 100 kbit/s.

The main objective is to investigate the side-effects of anti-inflammatory drugs. Therefore, dedicated biosensors are designed to evaluate the toxicity of drugs, the drug level, the glucose level, the ATP (Adenosine Tri-Phosphate) level, as well as the temperature and the pH level.

Implantable Bio-Monitoring System

Objectives
- Detection of different drugs
- Measurement of pH and temperature
- Detection of different endogenous compounds

Challenges
- Size and weight to be implantable
- Low coupling factor due to distance and to tissue

Swiss SNF Sinergia Project

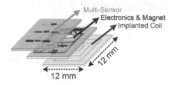

Conceptual design of battery-less implantable multiple sensor system

E. Kilinc, C. Baj-Rossi, S. Ghoreishizadeh, S. Riario, Fr. Stradolini,.C. Boero, G. De Micheli, F. Maloberti, S. Carrara, C. Dehollain, "A System for Wireless Power Transfer and Data Communication of Long-Term Bio-Monitoring", IEEE Sensors Journal, vol. 15, no. 11, Nov. 2015, pp. 6559-6569.

The most important challenge at system level is linked to remote power. The main reason is that the inductive coupling factor k is low between the primary coil of the external base station and the secondary coil of this implant and this coupling factor varies with time as the mouse moves in the cage.

The biochip sensor node is constituted by three layers (Fig. 4 of the cited article): (i) a first layer inside the body but close to the surface for the implanted secondary coil, (ii) a second layer for the CMOS chip and for the magnet which will be used to localize the mouse in the laboratory cage, (iii) a third layers with the sensors to measure different biological parameters.

This commercial ceramic thermistor (100K6MCD1 from BetaTHERM) has been selected due to its small dimensions, which are compatible for the implementation in the brown adipose tissue of a mouse, and to its high sensitivity. This thermistor is used to study its metabolism.

Thermistor Response Curve

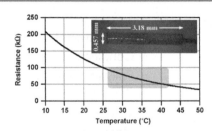

100K6MCD1, BetaTHERMSensors

- Metabolism study through brown adipose tissue temperature
- Target temperature range: 27 to 42 Degree Celsius
- Target resolution: 0.05 to 0.1 Degree Celsius

M.A. Ghanad, M.M. Green, C. Dehollain, "A 15uW 5.5 kS/s Resistive Sensor Read-Out Circuit with 7.6 ENOB", IEEE Transactions on Circuits and Systems I, vol. 61, no. 12, Dec. 2014, pp. 3321-3329.

Two thermistors are implemented in the mouse: a first one in the brown adipose tissue and a second one in another tissue of the mouse in order to measure the temperature difference between them.

The range of interest is between 27 degree Celsius and 42 degree Celsius which correspond to a variation in resistance from 91 to 48 KOhm (Fig. 1 of the cited article). In first approximation, a resolution of 0.05 degree Celsius corresponds to 108 Ohm.

19

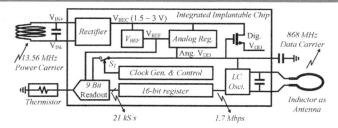

Low-power Implantable Chip

- □ High efficiency semi-active rectifier
- □ Time-domain resistance to digital converter
- □ Time interleaved sensor readout and data transmission

M. A. Ghanad, M. M. Green, and C. Dehollain, "A Remotely Powered Implantable IC for Recording Mouse Local Temperature with ±0.09 °C Accuracy," IEEE A-SSCC 2013 Conference.

This schematic represents the block diagram of the implantable chip (Fig. 1 of the cited article). The sensor node is remotely powered by the external base station at 13.56 MHz through magnetic coupling. The measurements of the thermistor are digitalized and they are transmitted by an active transmitter through the carrier at 868 MHz thanks to On-Off Keying (OOK) amplitude modulation. The external inductor of the LC tank of the oscillator is also used as an antenna.

The architecture of the rectifier is a semi-active topology with a high power efficiency of 85 % for an input power of 58 uW. The Vref block corresponds to the bandgap circuit (Vref = 0.46 V). The Analog Reg. Block corresponds to the voltage regulator such that its output Ang. Vdd is equal to 1.3 V. This first supply voltage is used for the analog blocks of the sensor node. The supply voltage Dig. Vdd is equal to 1 V. This second supply voltage is applied to the digital blocks of the sensor node.

20

Local Temperature Sensing Chip

- □ Semi active rectifier with leakage current control.
- □ Time-domain sensor readout.
- □ Duty cycled free-running oscillator for data communication.

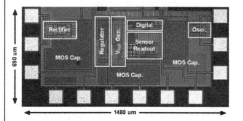

Implantable Chip Measurement	
Technology	0.18 um
Sensor type	Thermistor
Power	53 (uW)
V_{REC} Min.	1.5 (V)
Tran. Data Rate	1.7 (Mbps)
Sampling Rate	21 (kS/s)
Accuracy	±0.09 C°

M. A. Ghanad, M. M. Green, and C. Dehollain, "A Remotely Powered Implantable IC for Recording Mouse Local Temperature with ±0.09 °C Accuracy," IEEE A-SSCC 2013 Conference.

The integrated circuit has been designed in 0.18um CMOS technology. Its dimensions are equal to 690 um by 1480 um (Fig. 6 of the cited article). The inductor of the transmitter is external to the chip. The circuit operates under an input RF power of 53 uW. A minimum voltage of 1.5 V is required at the output of the rectifier.

The active LC transmitter operates at 1.7 Mbit/s. The sampling rate of the output signal of the sensor readout block is equal to 21 KSamples/s.

The commercial thermistor of Betatherm (100K6MCD1) has been chosen as temperature sensor. The temperature accuracy is equal to 0.09 Degree Celsius.

21 Time-domain Sensor Readout

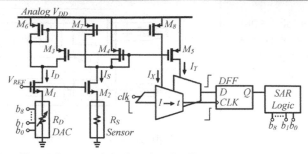

This schematic represents the readout circuit which has been designed in 0.18 CMOS technology (Fig. 4 of the cited article). The power consumption is equal to 17 uW at 21 KS/s. The resistive sensor is compared to a 9 bits DAC resistive ladder so that the difference between these two resistances is

- □ SAR algorithm tries to minimize the R_S-R_D
- □ Power consumption: 17 uW at 21 kS/s
- □ 9-bits (samples LSB twice)

M. A. Ghanad, M. M. Green, and C. Dehollain, "A Remotely Powered Implantable IC for Recording Mouse Local Temperature with ±0.09 °C Accuracy," IEEE A-SSCC 2013 Conference.

minimized thanks to the successive approximation register (SAR) algorithm. This difference is minimum when the drain current of transistor M1 is equal to the drain current of transistor M2. These two currents are compared thanks to two current to delay conversion stages and to the D Flip-Flop block (DFF block). The output of the DFF block is applied to the SAR logic block. The outputs of the SAR logic block control the 9 bits DAC resistive ladder.

22 Implemented Data Transmitter

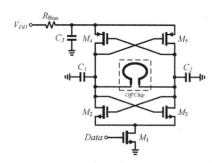

This schematic represents the data transmitter of the sensor node which operates at a carrier frequency of 868 MHz (Fig. 9 of the cited article). The On-Off Keying (OOK) amplitude modulation is selected. The transmitter is turned on and off thanks to the data signal applied to the gate of transistor M1. The total power

Spec.	Value
Center freq.	868 MHz
Modulation	OOK
Power	144 uW
Supply	1.3 V
Data-rate	1.7 Mbps
Turn on time	< 10 ns
Freq. drifts	< 1 MHz
Technology	0.18 um

- □ LC cross-coupled pair oscillator
- □ Resistive loss of the LC tank compensated by the negative resistances of the cross-coupled PMOS pair and of the cross-coupled NMOS pair.

M. A. Ghanad, M.M. Green, C. Dehollain, "A 30 uW Remotely Powered Local Temperature Monitoring Implantable System", IEEE Transactions on Biomedical Circuits and Systems, vol. 11, no. 1, pp. 54-63, Feb. 2017.

consumption is equal to 144 uW under a supply voltage of 1.3 V. The data rate is equal to 1.7 Mbit/s. The turn-on time of the transmitter is lower than 10 ns. The carrier frequency is determined by the two capacitors C1 and C2 and by the off-chip inductive antenna of the LC tank. A single-turn coil constitutes the antenna. Its diameter is equal to 1.1 cm. The maximum frequency drift of the centered frequency of the transmitter due to the technological variation of these two capacitors and of the off-chip inductor is equal to 1 MHz.

23   Wireless Power and Data Transfer for Intracranial Epilepsy Monitoring

The RFIC group of EPFL works in the domain of wireless power and data transfer for intracranial epilepsy monitoring since year 2007 thanks to the Swiss National Foundation (SNF) financial support and to the EPFL financial support.

The research work of the RFIC group is currently focused on the epilepsy monitoring. The next step will be to stimulate a specific portion of the brain as soon as an epilepsy seizure starts in order to try to decrease drastically its effect on the patient.

A book entitled *Wireless power transfer and data communication for neural implants. Case Study: epilepsy monitoring* has been published in year 2017 by Springer (authors: G. Yilmaz and C. Dehollain).

Dr. Mithat Silay and Dr. Gürkan Yilmaz have got their PhD thesis respectively in years 2012 and 2014. Mr. Kerim Türe has started his PhD thesis in year 2015 and it is planned that he will finish it in year 2019.

24 Drawbacks of Intracranial Neural Implants

More than 50 millions of people are affected worldwide by epileptic seizures. Long term medical treatments through dedicated drugs are efficient for around 70% of these patients.

That means that these medical drugs are not appropriate for 30%

- ☐ Risk of Cerebrospinal Fluid (CSF) leakage
- ☐ Risk of CSF infection
- ☐ Reduced patient comfort
- ☐ Short monitoring period

of the epileptic patients. In such cases, one solution can be performing a surgery to destroy the part of the brain which causes the seizures. Another alternative can be stimulating electrically this section of the brain to decrease the levels of the seizures.

For these two solutions, electrodes are used to determine the spot of the brain from which starts the seizure. Therefore, the patient goes to the hos-

pital so that the medical doctor takes off one part of the scalp and of the bone of the skull to put on the surface of the brain a network of electrodes which are connected through small cables to an external base station. During this stay at the hospital, there is a risk of Cerebraspinal Fluid (CSF) Leakage and the CSF can be infected.

25

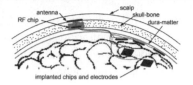

Wireless Power and Data Transfer System

The solution proposed by the RFIC group of EPFL is to perform a small burr hole in the bone of the skull. An implanted base station, constituted by a RF CMOS chip and an antenna (coil), is inserted in this hole (Fig. 1 of the cited article). Then, this hole is closed by the medical doctor and the patient can go back at home.

- Wireless Power Transfer by Magnetic Coupling at 8 MHz
- 30 % total power efficiency from 1cm for a DC delivered output power of 10 mW
- Implanted coil size < 15 x 15 mm2
- Polymer based packaging and modeling of the packaging
- At least 1 month of successful in-vitro operation

G. Yilmaz and C. Dehollain, "Wireless Communication and Power Transfer System for Intracranial Neural Recording Applications", Proceedings of the NEWCAS Conference, June 2014, 4 pages length.

The implanted base station is remotely powered by an external base station through magnetic coupling and it exchanges data with it. These data corresponds to the output signals of the read-out circuit connected to the electrodes. The read-out circuit is connected to the internal base station through small cables.

The frequency for remote power is equal to 8 MHz. The overall power efficiency is equal to 30% at a distance of 1 cm for a DC output power of 10 mW. The area of the implanted coil is equal to 1.5 cm * 1.5 cm. The internal base station has been packaged. Measurements in-vitro and in-vivo have been performed successfully.

26

Power and Data for Epilepsy Monitoring

The first solution uses a single frequency approach for which the same carrier signal at 8.3 MHz is used for magnetic remote powering and uplink data communication (from the internal to the external base stations). The total power efficiency in-vitro is equal to 33 % at a distance of 1 cm and for a delivered output DC power of 10 mW. Uplink data communication is performed by the modulation of the capacitor which is connected to the implanted coil of the transformer. A maximum data rate of 1 Mbit/s is achieved in-vitro.

- **Wireless Data Communication (single frequency approach)**
  - Remote power through magnetic coupling at 8.3 MHz
  - 33% total power efficiency in *in-vitro* experiments
  - Load modulation through the control of the implanted capacitor
  - 1 Mbit/s for uplink data communication
- **Wireless Data Communication (dual frequency approach)**
  - Remote power through magnetic coupling at 8 MHz
  - 30% total power efficiency in *in-vivo* experiments
  - Active transmitter at 416 MHz with OOK amplitude modulation
  - 12 Mbit/s for uplink data communication

G. Yilmaz, O. Atasoy, C. Dehollain, "Wireless Energy and Data Transfer for In-Vivo Epileptic Focus Localization", IEEE Sensors Journal, vol. 13, no. 11, Nov. 2013, pp. 4172–4179.
K. Türe, R. Ranjandish, G. Yilmaz, S. Seiler, H.R. Widmer, A. Schmid, F. Maloberti, C. Dehollain, "Power/Data Platform for High Data Rate in Implanted Neural Monitoring System", IEEE BIOCAS Conference, Oct. 2017, pp. 96-99.

The second solution uses a dual frequency approach for which the magnetic remote power and the uplink data communication are performed respectively at 8 MHz and at 416 MHz. The total power efficiency in-vivo is equal to 30 % at a distance of 1 cm and for a delivered output DC power of 10 mW. Uplink data communication is performed by an active transmitter (On-Off Keying amplitude modulation). A maximum data rate of 12 Mbit/s is achieved in-vivo.

27   Far-Field Remotely Powered Wireless Sensor System

The RFIC group of EPFL is active in the domain of far-field remotely powered wireless sensor systems since year 2000 thanks to the Swiss National Foundation (SNF) financial support, to the EPFL financial support and to the European Commission financial support.

The book "Design and optimization of passive UHF RFID systems" has been published in year 2007 by Springer (authors: J-P. Curty, M. Declercq, C. Dehollain and N. Joehl).

Dr. Jari-Pascal Curty, Dr. Nicolas Pillin, Dr. Onur Kazanc and Dr. Kerem Kapucu have got their PhD thesis respectively in 2006, 2010, 2014 and 2015. Firstly, J-P. Curty has focused his research on the maximization of the power transfer efficiency from the base station to the tag. Secondly, N. Pillin has added to the tag a memory to store information. Thirdly, O. Kazanc has co-designed the antenna of the tag and the rectifier to maximize the power transfer. Fourthly, K. Kapucu has added to the tag a CMOS low power read-out circuit dedicated to a capacitive sensor.

28   Adaptive Impedance Matching

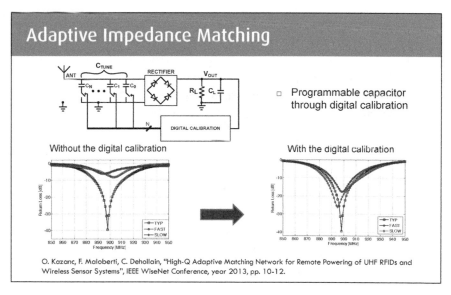

O. Kazanc, F. Maloberti, C. Dehollain, "High-Q Adaptive Matching Network for Remote Powering of UHF RFIDs and Wireless Sensor Systems", IEEE WiseNet Conference, year 2013, pp. 10-12.

A bank of capacitors is connected between the antenna and the rectifier of the tag (Fig. 1 of the cited article). The maximum transfer of power is obtained when the impedances of the bank of capacitors and of the antenna are complex conjugate. The input impedance of the rectifier depends on the temperature and on the CMOS process variations. Therefore, a programmable capacitor is implemented thanks to the bank of capacitors.

The system has been designed in 0.18 um CMOS technology for an input available power of 40 uW and an operating frequency of 900 MHz. The return loss (modulus in dB of the reflection coefficient) is represented with respect to the frequency for the nominal CMOS process parameters (TYP case) and for the corner CMOS process parameters (FAST and SLOW cases). The two graphs (Fig. 3 and Fig. 6 of the cited article) show that the return loss is improved at the operating frequency for the FAST and SLOW cases thanks this technique.

29

CMOS Differential Rectifier

The differential input architecture of Kotani has been selected due to its high power conversion efficiency (Fig. 2 of the cited article). The circuit has been optimized for an available input power of 40 uW, an operating frequency of 900 MHz and an output impedance of 1.2 nF in parallel with 180 kOhms.

Performance	Value
Wireless Remote Power	900 MHz
CMOS Technology	0.18 um
Available Input Power	40 uW
Conversion Power Efficiency in TYP Case	58 %
Architecture	From Kotani
Output Impedance	Storage capacitor: 1.2 nF Resistive Load: 180 KOhms

O. Kazanc, F. Maloberti, C. Dehollain, "High-Q Adaptive Matching Network for Remote Powering of UHF RFIDs and Wireless Sensor Systems", IEEE WiseNet Conference, year 2013, pp. 10-12.

The optimized dimensions of the NMOS transistors, of the PMOS transistors and of the Metal Insulator Metal (MIM) capacitors are respectively equal to W/L = 16 um / 0.18 um, W/L = 32 um / 0.18 um, and 1 pF. A conversion power efficiency of 58 % is obtained in simulation through Spectre circuit simulator by using the CMOS process parameters (TYP case).

30

Passive UHF RFID Tag

The architecture of a RFID passive tag has been selected for the European Marie-Curie FlexSmell project and the Swiss Hasler project to monitor the quality of the food through a humidity capacitive sensor. The sensor node is placed on the package of the food and it sends its measurements to a base station. The selected operating frequency band for remote power and uplink data communication is centered at 866 MHz. It has been selected due to the targeted distance (1.5 m) and its better performances, than at 13.56 MHz, with respect to misalignment between the antenna of the base station and the antenna of the tag.

The sensor node (Fig. 1 of the cited article) is

- □ **Capacitive Humidity Sensor**
- □ **Back-Scattering Wireless Com.**
- □ **Remote Powering Link**
  - □ Base Station
  - □ Base Station Antenna
  - □ Tag Antenna
  - □ Rectifier
- □ **Power Management Circuit**
  - □ Supply Voltage Generation
    - ■ Low Drop-Out Voltage Regulator
    - ■ Bandgap Reference
  - □ Current Reference

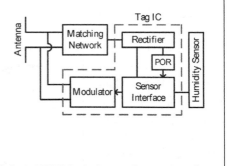

K. Kapucu, J. L. Merino Panades, C. Dehollain, "Design of a Passive UHF RFID Tag for Capacitive Sensor Applications", PRIME Conference, year 2013, pp. 213–216.

constituted by the antenna, the impedance matching network, the humidity sensor, the CMOS tag (rectifier, Power-On Reset circuit: POR, sensor interface, backscattering modulator, low drop-out voltage regulator, bandgap reference circuit, current reference circuit). The design of the base station is performed in discrete elements.

31 — Base Station and Tag Antennas

A patch antenna topology has been selected for the base station. It is directly impedance matched to the 50 Ohm coaxial cable. The maximum far-field gain is equal to 3.6 dBi at 866 MHz. The modulus of the reflection coefficient is equal to – 24 dB at 866 MHz. The dimensions are equal to 11cm * 8 cm.

□ **Base Station Antenna**
  ▫ Patch
  ▫ Matched to 50 Ohm coaxial cable
  ▫ Maximum far-field gain = 3.6 dBi at 866 MHz
  ▫ S11 = -24 dB at 866 MHz
  ▫ Dimensions: 11 cm * 9 cm

□ **Tag Antenna**
  ▫ Inductively Coupled Meandered Dipole
  ▫ Matched to chip impedance in TYP case
  ▫ Impedance at 866 MHz: 30 Ohms + j*188 Ohms
  ▫ Maximum far-field gain = 1.2 dBi at 866 MHz
  ▫ Dimensions: 8.3 cm * 2 cm

K. Kapucu and C. Dehollain, "Remote Powering Link for a Passive UHF RFID Tag for Capacitive Sensor Applications", ICECS 2013 conference, year 2013, pp. 823-826.

A differential inductively coupled meandered dipole antenna topology (Fig. 2 of the cited article) has been selected for the tag. In simulation, the antenna is directly matched to the input impedance of the differential input rectifier at 866 MHz by using the 0.18 um CMOS process parameters (TYP case). The impedance of the antenna at 866 MHz is equal to 30 Ohm + j* 188 Ohm. The maximum far-field gain is equal to 1.2 dBi at 866 MHz. The dimensions are equal to 8.3 cm * 2 cm.

The measured far-field radiation patterns of the base station antenna and the tag antenna are represented (Fig. 4 of the cited article).

32 — CMOS Differential Rectifier

The input of the rectifier works in differential mode to fit with the differential tag antenna. The topology of Kotani (Fig. 2a of the cited article) is selected to obtain high power conversion efficiency (PCE). The number of stages is equal to 3. The tag antenna and the CMOS rectifier are co-designed by using the 0.18 um CMOS process parameters (TYP case) so that the input impedance of the rectifier (30 Ohm - j* 188 Ohm) is the complex conjugate of the impedance of the antenna at the operating frequency of 866 MHz.

□ RF differential input signal
□ Kotani's topology
□ 0.18 um CMOS technology
□ Input Impedance at 866 MHz: 30 Ohms - j*188 Ohms
□ Matched at 866 MHz to tag antenna in TYP case
□ 1 mW output power and output resistive load of 5 Kohm
  ▫ Measured power conversion efficiency: 65%

K. Kapucu and C. Dehollain, "A Passive UHF RFID System with a Low-Power Capacitive Sensor Interface", IEEE RFID Technologies and Applications (RFID-TA) Conference, year 2014, pp. 301-305.

The measured PCE is equal to 65 % for a delivered output power of 1 mW and an output resistive load of 5 KOhm.

33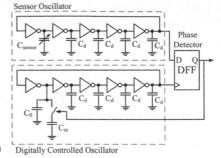

Low Power Sensor Interface

□ **Capacitive sensor**
  □ Printed humidity sensor
□ **Sensor readout**
  □ 0.18 um CMOS
  □ Supply voltage
    ■ Down to 0.8 V
  □ Low power
    ■ 12 uW at 0.8 V
□ **Distance range: 4 m**
  □ 3.3 W EIRP from the base station

K. Kapucu and C. Dehollain, "A Passive UHF RFID System with a Low-Power Capacitive Sensor Interface", IEEE RFID Technologies and Applications (RFID-TA) Conference, year 2014, pp. 301-305.

This sensor interface is dedicated to a printed capacitive sensor to measure the humidity. This interface (Fig. 2b of the cited article) is constituted by two ring oscillators, a phase detector and a CMOS switch designed in 0.18 um CMOS technology. The phase of the two ring oscillators are compared thanks to the phase detector. The switch is controlled by the output of the phase detector. The value of the capacitive sensor is equal to the sum of the fixed capacitor and of the time average value of the switched capacitor.

The sensor interface can operate down to a supply voltage of 0.8 V in 0.18 um CMOS technology. Its power consumption is equal to 12 uW. The maximum operating range is equal to 4 m when the base station sends to the sensor node 3.3 W Effective Isotropic Radiated Power (EIRP). This level of power corresponds to the maximum authorized value at a frequency of 866 MHz.

PART 2

34 PASSIVE TRANSMITTERS THANKS TO BACKSCATTERING DATA COMMUNICATION

Part 2 is dedicated to backscattering data communication from the sensor node to the base station. Firstly, the basic principle of backscattering data communication will be explained. Secondly, a model for the maximum distance range between the sensor node and the base station will be proposed and compared to measurements achieved at 900 MHz and at 2.4 GHz. Thirdly, a summary with respect to the radio regulations will be performed. Finally, an example of a passive memory tag will be given for which a dual frequency approach is mandatory to fulfill the system specifications (data rate, distance range, etc).

35

Backscattering Data Communication

□ The RF carrier is generated in the base station also called interrogator or reader
□ The power consumption of the tag is low because it does not include a RF oscillator to generate the carrier
□ The tag, also called transponder, modulates and reradiates the RF carrier that is coming from the interrogator when there is a high impedance mismatch between the tag antenna and the tag.

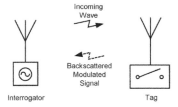

C. Dehollain, M. Declercq, N. Joehl, J.P. Curty, "A Global Survey on Short Range Low Power Wireless Data Transmission Architectures for ISM Applications", IEEE CAS Conference, year 2001, Vol. 1, pp. 117-126.

I n the case of a classical transmitter, the carrier is generated by a RF oscillator which is included in the transmitter of the tag. To decrease the power consumption of the tag, the idea is to generate the carrier in the base station also called interrogator or reader (Fig. 6 of the cited article).

The tag modulates and reradiates the incoming RF carrier signal thanks to a switch which is controlled by the data to be sent to the interrogator.

This technique is called backscattering modulation because the characteristics of the signal (amplitude or phase) which is backscattered by the tag in the direction of the interrogator depend on the data to be transmitted.

36

Implementation of the Data Communication

If Data = Bit "1"
- Zin = R' is mismatched to the tag antenna
- The RF incoming signal is reflected to the interrogator

If Data = Bit "0"
- Zin = R is matched to the tag antenna
- The RF incoming signal is absorbed by the tag

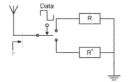

f_0: carrier and f_D: data

C. Dehollain, M. Declercq, N. Joehl, J.P. Curty, "A Global Survey on Short Range Low Power Wireless Data Transmission Architectures for ISM Applications", IEEE CAS Conference, year 2001, Vol. 1, pp. 117-126.

B ackscattering modulation is obtained by modulating the input impedance of the tag with the data stream to be transmitted (Fig. 7 of the cited article). The assumption is done that the data consist in a binary signal.

It is assumed that the impedance of the tag antenna is equivalent to a resistor R at the desired RF frequency (e.g. 868 MHz) and that the resistor Ron of the commutator is negligible.

The resistor R' is chosen much smaller or much larger than the resistor R. The amplitude of the signal reflected by the tag is switched between two values when the commutator is connected to R or to R'. This amplitude variation corresponds to an On Off Keying (OOK) amplitude modulation.

The frequencies of the RF and data signals are respectively denoted by f0 and fD. The components of the backscattered signal sent by the tag are located on both sides of f0. The two components at the fundamental frequency are represented in dashed line on both sides of the carrier frequency.

37 IF Backscattering Data Communication

The backscattered signal is demodulated by the receiver of the interrogator. The phase noise of the carrier signal (often called Local Oscillator signal: LO) generated by the interrogator has an effect on the performances of the receiver because the same carrier is used to down-convert the backscattered signal.

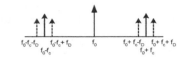

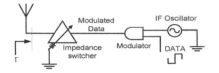

f_0: carrier frequency
IF: Intermediate Frequency
f_c: IF frequency
f_D: data frequency

C. Dehollain, M. Declercq, N. Joehl, J.P. Curty, "A Global Survey on Short Range Low Power Wireless Data Transmission Architectures for ISM Applications", IEEE CAS Conference, year 2001, Vol. 1, pp. 117-126.

The effect of the phase noise can be reduced if the frequency shift between the components of the backscattered signal and the carrier is increased. The solution (Fig. 12 of the cited article) is that the data modulate an IF (Intermediate Frequency) oscillator by using a simple mixer (logic gate). The resulting signal, called "Modulated data", corresponds to the control signal of the impedance switcher.

This architecture provides an efficient way to increase the sensitivity as well as a multi-channel coding ability. A compromise has to be made between the power consumption of the IF oscillator and the sensitivity of the system in order to choose the value of the IF frequency.

38 Modulation Types

Binary data are often used to modulate the carrier signal sent to the tag by the interrogator. These data modulate the input impedance of the tag as well as the corresponding reflection coefficient of the tag with respect to its antenna.

The Binary Amplitude Shift Keying (BASK) modulation corresponds to the modulation of the

- ■ The reflection coefficient at the tag-antenna interface can vary in
  - ■ Amplitude
  - ■ Phase
- ■ Two basic binary modulation types are often used
  - ■ Binary Amplitude Shift Keying (BASK)
  - ■ Binary Phase Shift Keying (BPSK)
- ■ Comparison of BASK and BPSK modulations through
  - ■ Power available for remote power and for uplink data communication
  - ■ Bit Error Rate (BER) for a given energy per bit over noise ratio at the input
  - ■ Architecture of the receiver of the interrogator

J.P. Curty, M. Declercq, C. Dehollain, N. Joehl, "Design and optimization of passive UHF RFID systems", Springer, year 2007

reflection coefficient in amplitude. The Binary Phase Shift Keying (BPSK) modulation corresponds to the modulation of the reflection coefficient in phase. To compare BASK modulation to BPSK modulation, it is necessary to study the effect of the modulation on the remote power operation, when the same car-

rier is used for remote power and uplink data communication, to estimate the corresponding Bit Error Rate (BER) for a given Energy per bit over Noise ratio at the input of the receiver, and to determine the corresponding architecture of the receiver of the interrogator.

39 — Read Range of Far Field RFID Systems

This figure (Fig. 1 of the cited article) represents the reader and the tag. The goal of this study is to estimate the distance range for data communication as a function of different parameters at system level. It will be shown that the phase noise of the local oscillator and the coupling factor between the two antennas of the reader have an impact on it.

A model is derived to estimate the distance range of passive far-field RFID systems

The phase noise of the local oscillator in the reader and the coupling factor between the two antennas of the reader have an important impact

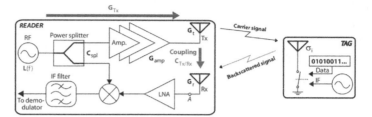

N. Pillin, C. Dehollaïn, M. Declercq, «Read Range Limitation in IF-Based Far-Field RFID using ASK Backscatter Modulation», PRIME Conference, year 2009, pp. 348-351.

The carrier signal is generated by the Local Oscillator (LO) of the reader. The power splitter divides this signal in two paths: a first one to the antenna for the transmission, a second one to the mixer for down-conversion of the backscattered signal. The quality of the isolation between the antenna for transmission and the one for reception is quantified by the coupling factor.

Backscattering modulation is achieved thanks to a switch which is controlled by a modulated signal. This modulated signal corresponds to the data which are themselves modulated by an Intermediate Frequency (IF) subcarrier signal generated by an oscillator.

40 — Effective Radar Cross Section

Due to the backscattering operation, the Radar Cross Section (RCS) of the tag antenna depends on the data to be transmitted from the tag to the reader. More precisely, the RCS is equal to one value when the data corresponds to a bit equal to "1" and to another one when the bit is equal to "0".

Due to the backscattering operation, the antenna of the tag switches between two different Radar Cross Section called RCS: σ_1 and σ_2

Effective tag antenna RCS is expressed by

$$\sigma_i = \left(\frac{\sqrt{\sigma_1} - \sqrt{\sigma_2}}{2} \right)^2$$

N. Pillin, C. Dehollaïn, M. Declercq, «Read Range Limitation in IF-Based Far-Field RFID using ASK Backscatter Modulation», PRIME Conference, year 2009, pp. 348-351.

The RCS difference is quantified by the effective tag antenna radar cross section. It is defined by the given equation. It can be mentioned that the de-modulation of the backscattered signal by the receiver of the reader is easy if the effective radar cross section is high.

41 Estimation of the Maximum Distance Range

The estimation of the maximum distance range in the case of backscattering data communication is given by this equation.

$$d_{max} = \sqrt[4]{\frac{G_t G_r \alpha \sigma_i}{C_{Tx/Rx} N(f_{SB}, B) SNR_{min}} \frac{\lambda^2}{(4\pi)^3}}$$

- Gt: gain of the antenna of the reader for transmission
- Gr: gain of the antenna of the reader for reception
- CTx/Rx: coupling factor between the two antennas of the reader
- N (fsb, B): integral of the phase noise of the carrier in the bandwidth B
- SNRmin: minimum signal over noise ratio at the input of the reader

N. Pillin, C. Dehollain, M. Declercq, «Read Range Limitation in IF-Based Far-Field RFID using ASK Backscatter Modulation», PRIME Conference, year 2009, pp. 348-351.

The expression of the numerator is logical as the distance range increases if the gains of the two antennas of the reader increase. The same is true if the effective radar cross section of the tag antenna or if the wavelength of the carrier generated by the reader increases.

The expression of the denominator is also logical as the distance range decreases if the coupling factor between the two antennas of the reader increases. The same is true if the phase noise of the carrier or if the minimum signal over noise ratio at the input of the reader increases.

42 Parameters of the Tag and of the Reader

Two tags and two readers have been built by using discrete electronic components to validate the model at 900 MHz and at 2.45 GHz. The schematic of the tag is represented by the figure (Fig. 3 of the cited article).

The main parameters of the two tags are summarized in this table. The impedance

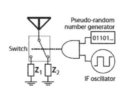

	Parameters at 0.9 GHz	Parameters at 2.45 GHz	Unit
f_{IF}	10	10	MHz
G_t	2	7	dBi
G_r	2	7	dBi
$C_{Tx/Rx}$	-23	-27	dB
Z_1	50	50	Ohm
Z_2	0	infinity	Ohm
σ_i	64.4	17.4	cm^2
α	-7	-7	dB

N. Pillin, C. Dehollain, M. Declercq, «Read Range Limitation in IF-Based Far-Field RFID using ASK Backscatter Modulation», PRIME Conference, year 2009, pp. 348-351.

of the tag antenna for each prototype is equal to 50 Ohms at the operating frequency. Due to the backscattering operation, the impedance connected to the tag is either equal to 50 Ohm or to a short circuit at 900 MHz whereas it is equal to 50 Ohm or to an open circuit at 2.45 GHz.

With respect to the reader, the coupling factor between the antenna for transmission and the one for reception as well as the gains of these two antennas are also mentioned in this table.

43

Measurements compared to Model

Measurements were achieved on a laboratory reader/tag system at 900 MHz and 2.45 GHz to verify the model.

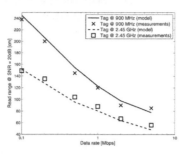

N. Pillin, C. Dehollain, M. Declercq, «Read Range Limitation in IF-Based Far-Field RFID using ASK Backscatter Modulation», PRIME Conference, year 2009, pp. 348-351.

It is recalled that the backscattering data communication from the tag to the reader operates in far-field. Measurements have been performed at 900 MHz and at 2.45 GHz by using the two discrete components readers and tags. This figure (Fig. 4 of the cited article) represents the distance range as a function of the data rate. The targeted minimum signal over noise ratio at the input of the reader is fixed to 20 dB. The dashed line corresponds to the model at 2.45 GHz and the plain line to the one at 900 MHz.

These results show a good agreement between the measurements and the model at 900 MHz and at 2.45 GHz.

44

Radio Regulations

□ Some regulations are fine for data communication (large frequency bandwidth)
□ Others for wireless power transfer (high transmitted power)

Frequency Band [GHz]	Maximum Transmitted Power [W]	Frequency Bandwidth [MHz]	Suitable for
2.4 to 2.4835	0.01 ERP	83.5	Data Communication
0.8656 to 0.8676	2 W ERP[1]	2	Remote Powering
2.446 to 2.454	4 W EIRP[2]	8	Data Communication and Remote Powering

([1]~3.3 W EIRP, [2]for indoor use)

There are radio regulations to define the value of the carrier, the frequency bandwidth around the carrier as well as the maximum allowed transmitted power expressed in «Effective Radiated Power» (ERP) or in «Effective Isotropic Radiated Power» (EIRP). The conversion from ERP to EIRP is such that 1 W ERP = 1.65 W EIRP.

The table summarizes the parameters of three main frequency bandwidths used for far-field remote power and data communication. The first frequency band is suitable for data communication as the frequency bandwidth is high but the maximum allowed transmitted power is low. The second one is well adapted to remote power because the maximum allowed transmitted power is high. The last frequency band allows a good compromise between power and data communication as the maximum allowed power is high and the value of the frequency bandwidth can fit with a lot of sensor network applications.

45

Passive Memory Tag

User specifications
- High capacity non volatile memory to store multimedia content (video, music, etc)
- Minimum distance range: 0.3 m

Electronic specifications
- Read/write data rate: 10 Mbit/s
- Total power consumption (tag and non volatile memory): 10 mW
- CMOS tag manufactured in a standard 0.18 um CMOS process
- Rectifier power conversion efficiency: 60%

European IP
Project MINAMI

The concept of passive memory tag is defined by adding a non volatile FRAM memory to the tag to store information such as video, music, pictures, etc. It is possible to write in the FRAM by sending data from the reader and also to read its content thanks to data transmission from the tag to the reader. The minimum operating range between the reader and the tag is equal to 0.3 m in read and write modes.

The write and read operations need to be quick. Therefore, a data rate of minimum 10 Mbit/s is required. The maximum targeted total power consumption is equal to 10 mW to be compatible with the distance range and the radio regulations. Finally, the estimated rectifier power conversion efficiency is equal to 60 %.

46

Dual Frequency Passive Memory Tag

Friis propagation relation

$$P_{AV} = P_{EIRP}.G_R.\frac{\lambda^2}{(4\pi d)^2}$$

MANDATORY TO HAVE A DUAL FREQUENCY TAG: 868 MHz, 2.44 GHz

Data transmission
- Data Rate = 10 Mbit/s → Mini Bandwidth = 10 MHz
- Frequency for data transmission: 2.4 GHz to 2.485 GHz

Remote power
- Gain of the tag antenna Gr = 1
- Distance between the reader and the tag d = 0.3 m
- Power consumption: 10 mW
- Power conversion efficiency of the rectifier = 60% → Minimum Pav = 16.4 mW
- f = 2.4 GHz, Peirp = 4 W → Pav maxi = 4.05 mW: Not OK
- f = 868 MHz, Peirp = 3.3 W → Pav maxi = 25.5 mW: OK

In the case of far-field operation, the maximum RF active power at the input of the tag is given by the Friis propagation relation when a perfect impedance matching is fulfilled between the antenna and the tag. This active power depends on the effective isotropic radiated power sent by the reader, on the gain of the tag antenna, on the wavelength of the RF carrier and on the distance between the tag and the reader.

Due to the data rate of 10 Mbit/s and the distance range of 0.3 m, it is shown that a dual frequency approach is mandatory. The frequency band centered at 868 MHz is the appropriate choice for remote power whereas the one centered at 2.44 GHz is the good one for uplink and downlink data communication.

PART 3

47 REMOTE POWER FOR WIRELESS SENSOR NETWORKS

Part 3 is dedicated to remote power from the base station to the sensor node by using either electromagnetic coupling or magnetic coupling. The attenuation in air of the RF power sent by the base station to the sensor node will be estimated for different frequencies and distance range between the base station and the sensor node. The example of an implanted sensor node in a living mouse will be studied. Different architectures will be proposed for this application by using magnetic coupling.

48 Power by Electro-Magnetic Coupling

Firstly, the Friis propagation relation is expressed with respect to the effective isotropic radiated power. Secondly, this power is replaced by the transmitted power multiplied by the antenna gain of the transmitter of the base station (also called reader or interrogator). It is important to recall that the maximum available power is delivered to the sensor node (also called tag) when the impedance of the antenna of the sensor node is the complex conjugate of the input impedance of the sensor node at the operating frequency.

$$P_{AV} = P_{EIRP}.G_R.\frac{\lambda^2}{(4\pi d)^2}$$

$$P_{EIRP} = P_T.G_T$$

$$\Rightarrow \quad P_{AV} = P_T.G_T.G_R.\frac{\lambda^2}{(4\pi d)^2}$$

- ☐ The maximum available power P_{AV} in the case of impedance matching depends on
  - the transmitted power P_T
  - the antenna gain G_T of the transmitter
  - the antenna gain G_R of the tag
  - the frequency of the RF signal
  - the distance d between the two antennas

49

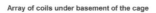

Remote Powering of an Implant

A sensor node is implanted in a mouse to measure different physiological parameters. The area of the cage and the distance between the external coils and the implant are respectively equal to 23 cm * 16 cm and to 3 cm. This sensor node is remotely powered by magnetic coupling. The

□ Low coupling factor of the inductive link due to the distance between the two coils and the limited dimensions of the implanted coil

□ The coupling factor k depends on the position of the mouse

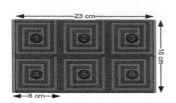

Array of coils under basement of the cage

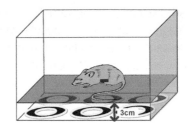

3D model of array of coils under basement

E.G. Kilinc, C. Dehollain, F. Maloberti, "Design and optimization of inductive power transmission for implantable sensor system", SM2ACD workshop, year 2010, pp. 1-5.

primary coils of the transformer are placed under the laboratory cage whereas the secondary coil is implanted in the mouse. A first solution consists of using an array of external coils (Fig. 7 of the cited article) because the mouse moves inside the cage.

50

Geometry of the Coils

The design parameters are determined by the dimensions of the mouse, by the application and by the process manufacturing of the coils. The Industrial Scientific Medical (ISM) frequency band centered at 13.56 MHz has been selected as the dimensions of the implanted coil are limited by the ones of the mouse.

DESIGN PARAMETERS LIMITED BY APPLICATION

Parameter	Value
Link operation frequency (f)	13.56 MHz
Distance between coils (d_{12})	30 mm
Tag coil outer diameter (d_{o2})	20 mm
Minimum width of conductor (w)	150 um
Minimum spacing between line (s)	150 um

OPTIMAL INDUCTIVE COIL DESIGNS

Parameter	Reader Coil	Tag Coil
Outer diameter (d_o)	80 mm	20 mm
Inner diameter (d_i)	10 mm	11 mm
Number of turns (n)	5	6
Width of conductor (w)	1 mm	250 um
Spacing between lines (s)	7.5 mm	600 um

E.G. Kilinc, C. Dehollain, F. Maloberti, "Design and optimization of inductive power transmission for implantable sensor system", SM2ACD workshop, year 2010, pp. 1-5.

With respect to the geometry of the two coils, it can be shown that a square shaped coil is preferable to a circular shaped one. That is the reason why the square shaped geometry has been selected for the two coils.

An analytical model of the transfer of power from the external coil to the implanted coil has been developed in the cited article and then implemented in Matlab in order to maximize it. This analytical model has been validated by using the CAD tool HFSS.

51

Comparison of the Two Types of Coupling

The aim of this study is to compare in air electro-magnetic coupling to magnetic coupling. Therefore, the attenuation due to the mouse tissues is not considered in this study. The assumption is made that impedance matching is achieved.

For electro-magnetic coupling, the gain anten-na of the base station and the gain antenna of the sensor node are equal to 1.

- Friis propagation relation: gain of the two antennas = 1
- Primary and secondary coils of SM2ACD 2010
- Path Loss = 10 log (P_{AVS} / P_T)
- Path Loss at 13.56 MHz = - 10 dB
- Pass Loss at 2.4 GHz = - 20 dB

$$\frac{P_{AVS}}{P_T} = \left(\frac{\lambda}{4\pi}\right)^2 \left(\frac{1}{d}\right)^2$$

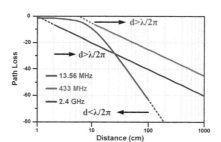

E.G. Kilinc, C. Dehollain, F. Maloberti, "Design and optimization of inductive power transmission for implantable sensor system", SM2ACD workshop, year 2010, pp. 1-5.

For magnetic coupling, the primary and second-ary coils of the transformer correspond to the ones of the cited article. They have been optimized to maximize the transfer of power at 13.56 MHz.

At a distance of 10 cm, the path loss at 13.56 MHz for magnetic coupling is lower than at 2.4 GHz for electro-magnetic coupling.

The wavelength in air at 433 MHz is equal to 66 cm. Therefore, a distance of 10 cm corresponds to the boundary between magnetic and electro-mag-netic coupling. That is the reason why it is difficult to compare at 10 cm the path loss at 433 MHz to the one at 13.56 MHz.

52

Solution 1: Fixed External Coils

The implanted sensor node is remotely powered by magnetic coupling at 13.56 MHz. Each powering coil of the array is connected to a Power Amplifier (PA) so that it provides power wirelessly to the sensor node (Fig. 2 of the cited article). A technique is proposed to localize the position of the mouse inside the

- **Cage with array of powering coils and magnetic field sensors**

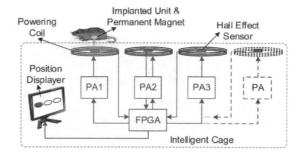

E.G. Kilinc, B. Canovas, F. Maloberti, C. Dehollain, "Intelligent cage for remotely powered freely moving animal telemetry systems", IEEE ISCAS Conference, year 2012, pp. 2207-2210.

cage thanks to the use of Hall effect sensors and of a small magnet respectively placed in the middle of the external coils and in the middle of the implanted coil. By this way, it is possible to turn on the minimum number of PAs to avoid cross-talk between the coils of the array and to minimize the power consumption of the PAs. An FPGA (Field Programmable gate Array) turns on and off the PAs thanks to an intelligent feedback mechanism.

53

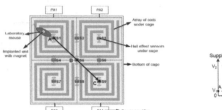

Power Management of the Power Amplifiers

- ☐ The mouse moves from A to B, then to C
- ☐ Smart power management of the Power Amplifiers
- ☐ Animal tracking

E.G. Kilinc, B. Canovas, F. Maloberti, C. Dehollain, "Intelligent cage for remotely powered freely moving animal telemetry systems", IEEE ISCAS Conference, year 2012, pp. 2207-2210.

An array of four external coils under the cage is considered in this figure (Fig. 3 of the cited article) but the methodology can be generalized to more coils. Hall effect sensors are placed in the middle of each coil (S1, S3, S7, S9) and at the boundary between the coils (S2, S4, S6, S8). The mouse contains a magnet. The output of the Hall effect sensor S1 (S9) is maximum when the mouse is at position A (at position C). Therefore, when the mouse moves from A to C, firstly PA1 is turned on when the mouse is close to A (Fig. 4 of the cited article). Then PA4 is also turned on when the mouse is close to B. Finally, PA1 is turned off when the mouse is close to C. Each PA is a class-E power amplifier to get a high power efficiency.

54

Solution 2: Moving External Coil

- ☐ Rails move at the maximum speed of 30 cm/s
- ☐ Faster than animal inside cage (~7 cm/s)

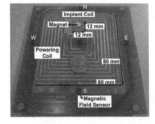

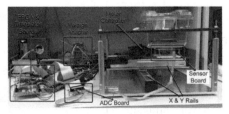

E.G. Kilinc, C. Dehollain, "Intelligent Remote Powering», EPO Patent 12180919.8, August 17, 2012 and PCT/EP2013/056611 Patent, August 13, 2013
E. G. Kilinc, G. Conus, C. Weber, B. Kawkabani, F. Maloberti, and C. Dehollain," A system for wireless power transfer of micro-systems in-vivo implantable in freely moving animals," IEEE Sensors Journal, vol. 14, pp. 522–531, Feb. 2014.

The external coil moves thanks to two rails (x-axis, y-axis). The position of this coil is controlled by a motor at a maximum speed of 30 cm/s which is higher than the one of the mouse which is equal to 7 m/s (Fig. 10 of the IEEE Sensors Journal paper). The motor is controlled by the output signals of the Hall sensors which are placed in the center and on the edges of the external coil (Fig. 9 of the IEEE Sensors Journal paper). The small magnet is placed in the center of the implant coil. The power efficiency of this second solution is higher than the one of the first approach as the alignment between the implant coil and the external coil is better thanks to the motor. The living mouse has been replaced by a freely moving robot with a magnet for in-vitro experiments. The maximum measured misalignment between the two coils is equal to 1 cm.

55

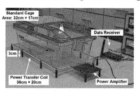

Solution 3: External Coil around the Cage

- ☐ Measurement of the temperature of the brown adipose tissue of the mouse to study its metabolism.
- ☐ Average DC power consumption of the implanted electronics: 30 uW.
- ☐ Power transfer efficiency between the external and the implanted coils:
  - ☐ 0.15% in the middle of the cage
  - ☐ 0.60 % at the edges of the cage

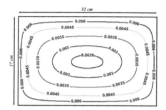

M. A. Ghanad, M.M. Green, C. Dehollain, "A 30 uW Remotely Powered Local Temperature Monitoring Implantable System", IEEE Transactions on Biomedical Circuits and Systems, vol. 11, no. 1, pp. 54-63, Feb. 2017.

The power transfer coil is placed around the cage at the height of 3 cm which corresponds to the optimum one when the mouse stands up (Fig. 10.a of the cited article). The designs of this coil and of the implanted one have been optimized at 13.56 MHz to maximize the transfer of power. The measured value, quality factor and number of turns of the external coil are equal respectively to 1.2 uH, 38 and 1. The shape of the implanted coil is a solenoid one with a measured value, quality factor, number of turns, diameter and height equal respectively to 4.6 uH, 78, 14, 1.4 cm and 1.4 cm.

The coupling factor k between the external coil and the implanted coil is represented on this graph (Fig. 10.b of the cited article). It depends on the position of the mouse. It is minimum in the middle of the cage and maximum at the edges of the cage. Therefore, the power transfer efficiency between the two coils varies from 0.15 % (in the middle of the cage) to 0.6 % (at the edges of the cage).

56

Conclusion

- ☐ Architectures of remotely powered wireless systems
  - ☐ Single frequency approach for data transfer and power transfer
  - ☐ Dual frequency approach for data transfer and power transfer

- ☐ Transfer of data
  - ☐ Passive transmitters: backscattering modulation, load modulation
  - ☐ Active transmitters
  - ☐ Low power radio frequency integrated circuits

- ☐ Transfer of power
  - ☐ Wireless energy sources: electromagnetic, magnetic, ultrasound, etc.
  - ☐ Antennas, transformer, electro-acoustic transducers
  - ☐ Power management for low power consumption

The architecture of the remotely powered system depends on the system specifications (distance range, data rate, power consumption, etc) as well as on the radio regulations. To get a simple architecture, it is better to use the same frequency band for remote power and data communication but it can be necessary to use two different frequency bands due to the specifications.

The transfer of data from the sensor node to the base station can be performed through a passive transmitter (generation of the RF carrier in the base station) or by an active transmitter (generation of the RF carrier in the sensor node). The choice be-

tween these two methods depends on the targeted power consumption of the transmitter and on the data rate.

The choice of the wireless energy source to transfer power from the base station to the sensor node depends on the system specifications (distance range, power consumption of the sensor node, etc).

CHAPTER 03

A System on Chip for Energy Harvesting and Wireless Power Transfer

Roberto La Rosa

Ultra Low Power ICs and Applications
Team Manager
STMicroelectronics

The talk entitled *A System on Chip for Energy Harvesting and Wireless Power Transfer* is motivated by the continuous development of IoT (Internet Of things) infrastructure and applications which is paving the way to advanced and innovative ideas and solutions, some of which are pushing to the limit the state-of-the-art technology. The increasing demand of WSN (Wireless Sensor Nodes) Nodes which need to be capable to collecting and sharing data wirelessly while often positioned in places hard to be reached and serviced, motivates engineers to look for innovative solutions of energy harvesting and wireless power transfer to allow battery-free sensor nodes. RF harvesting and wireless power transfer, due to the pervasiveness of RF energy, that can reach out of sight places, could be a key technology to power wirelessly IoT sensor devices, that in order to be ubiquitous, need to be wireless, maintenance free, battery free and low cost enough to be used almost anywhere. A System on Chip is presented to be used either as RF power receiver and as a Ultra Low Power high performance power management for multi source energy harvesting. Several different possible applications example are shown.

1 Presentation Outline

- **Motivation**
- **Introduction**
- **Overview**
- **System Description**
- **Design Notes**
- **Experimental Results**
- **Conclusions**

The talk break-down in seven differents subsection addressing the motivation, and introducing the context of an industrial product on Wireless Power Transfer.

This move to the core section of the talk providing an architecture overview and a system description.

Typical design and applications get discussed further.

Reaching a section to draw conclusion on Wireless Power Transfer Pros and cons.

2 Impact of Energy Harvesting and WPT on IoT

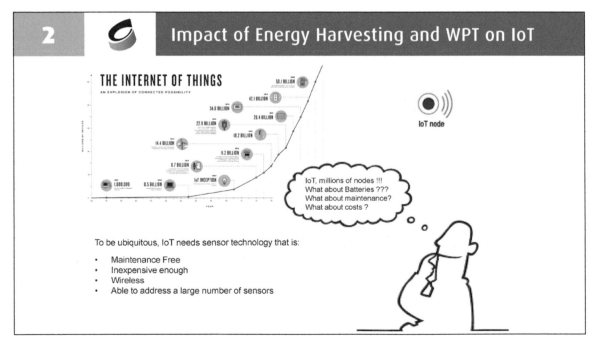

Market for IoT related devices is very promising with a forecast of 50 billions of "things" to be connected by 2020. Although very promising, there are still some consistent problems to be solved in order to render those technologies ubiquitous and convenient as it is still needed a mature sensor technology that is Wireless, Maintenance-Free, inexpensive enough and able to address a large number of sensors.

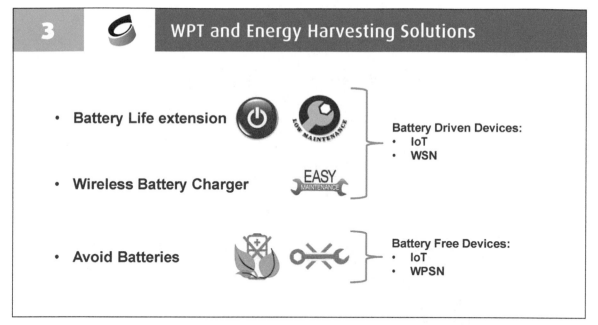

In order to relieve the maintenance issue in IoT and WSN (Wireless Sensor Networks) different strategies are possible depending on the kind of system. Some system might need the battery for the kind of sensing that is required while some others could be designed free of batteries.

In battery powered devices a good achievable improvement is to null standby power consumption. In fact while in standby, power management circuits are permanently on, consuming an unnecessary high percentage (50%-70%) of the battery charge. Nulling standby is a real saving which indeed provides either a battery life extension or a battery size reduction with an immediate benefit in terms of costs and system miniaturization.

Even nulling standby power consumption the nodes maintenance problem is only relieved but not truly solved since in most cases the nodes are often positioned in places difficult to reach. One of the main problems in WSN is battery life which poses a real problem in node maintenance, an operation often difficult and expensive to perform. For this reason it is needed a realistic and convenient way to ease the maintenance of battery-powered nodes through the implementation of an over-the-distance Wireless Battery Charger using RF (Radio Frequency) Energy Harvesting.

Where possible the node maintenance can be avoided with the ultimate solution of set-and-forget battery-free sensors which eliminate the limitation of system lifetime due to the presence of a battery, with the benefit of being non-disposable, more cost efficient and functional as long as power is delivered.

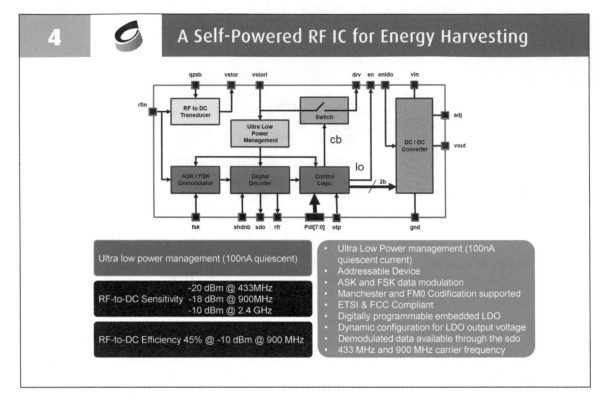

4 — A Self-Powered RF IC for Energy Harvesting

Ultra low power management (100nA quiescent)

RF-to-DC Sensitivity
-20 dBm @ 433MHz
-18 dBm @ 900MHz
-10 dBm @ 2.4 GHz

RF-to-DC Efficiency 45% @ -10 dBm @ 900 MHz

- Ultra Low Power management (100nA quiescent current)
- Addressable Device
- ASK and FSK data modulation
- Manchester and FM0 Codification supported
- ETSI & FCC Compliant
- Digitally programmable embedded LDO
- Dynamic configuration for LDO output voltage
- Demodulated data available through the sdo
- 433 MHz and 900 MHz carrier frequency

The proposed System on Chip is a self-powered RF IC for energy harvesting and RF wireless power transfer. This device integrates a high performance wide-bandwidth energy transducer (350 MHz – 2.4 GHz), with a sensitivity of –18.8 dBm at 900 MHz, a power efficiency of 45% at 900 MHz with an input power of –10 dBm and an accurate ASK / FSK demodulator with a modulation index as low as 10% and a sensitivity of -38 dBm.

The ultra low power management smartly manages the harvested energy and provides a regulated power supply to the internal circuitry of the system on chip as soon as the amount of energy is enough to guarantee the active phase of the system. It is made of a nano-power circuitry that performs a voltage sensing on the external storage capacitor to control the charging process with a current consumption as low as 75 nA, and of a micro-power

circuitry that mainly provides the regulated power supply to the internal circuitry of the system on chip.

The ASK receiver has been designed to detect a Mancester coded data stream with a minimum data rate of 62 kbit/s to allow a channel bandwidth occupation of around 250 KHz that is compliant with European regulation ETSI 300-200.1.

The FSK receiver has been designed to detect a Manchester coded data stream with a minimum data rate of 250 Kbit/s. The receiver supports both the 433 MHz and 915 MHz ISM bands.

The use of the Manchester coding is important to reduce both channel bandwidth occupation and system complexity. As well known, Manchester decoding can be performed with simple, low power, digital circuitry base on time-delay cells.

The IC integrates an adjustable and digital programmable LDO along with several other features.

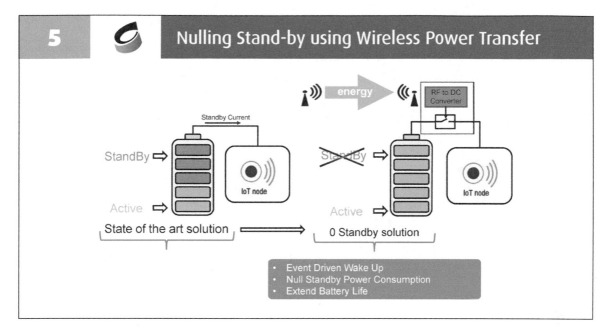

In order to increase energy efficiency in electrical appliances, autonomy in WSN (Wireless Sensor Network) nodes, IoT (Internet of Things) devices and self-powered sensors, it is necessary to reduce energy consumption as much as possible. While in standby, power management circuits are permanently on, consuming unnecessary energy. A different approach is introduced based on Radio Frequency to electrical energy transduction and wireless power transfer. **The proposed solution goes beyond the well-known concept of standby as it instead applies to electric appliances that are off.** In order to achieve this an RF powered receiver silicon IC (integrated circuit) for remotely controlled systems is used. This includes an RF-to-DC energy converter specifically designed with a sensitivity of -18.8 dBm, which allows an operating distance of up 8 meters at 900 MHz with a transmitting power of 0.5 Watt in free space. The basic idea is to efficiently convert a small amount of RF energy into sufficient electrical energy to switch on any electrical appliance. One of the main issues is that the RF energy received cannot be very high for several reasons, such as distance attenuation and limited transmitting power of the remote control unit. Therefore to reach the target distance of 8 meters, the transducer must be very efficient and carefully designed for power sensitivity and the system needs an ultra low power management with a quiescent current on 75 nA to allow the self-powering operation with a current as low as 1uA at sensitivity.

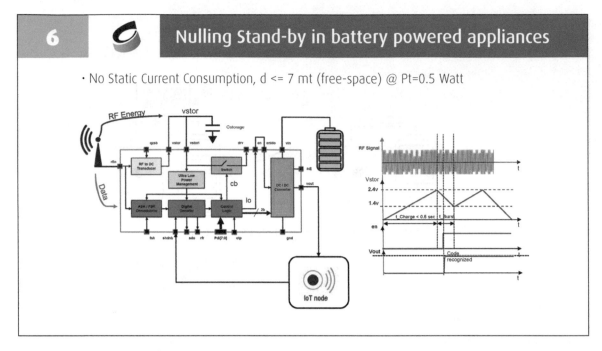

6 — **Nulling Stand-by in battery powered appliances**

· No Static Current Consumption, d <= 7 mt (free-space) @ Pt=0.5 Watt

The basic idea is to efficiently convert a small amount of RF energy into sufficient electrical energy to switch on any electrical appliance.

The appliance, such as an IoT device, is supplied by a DC/DC converter which is enabled by the digital decoder through the control logic, once the RF-to-DC transducer has provided enough power to supply those circuits.

As shown in the figure on the right, when the system is off, the Vout regulated voltage of the DC/DC converter is zero, all the other units of the IoT device are completely off and the battery does not consume any power.

To wake up and bias the system an RF signal is transmitted by a remote control unit with enough energy to turn the system on (0.5 Watt from a distance of 8 meters at 900 MHz in free space). Once the RF antenna catches the energy of the RF signal, the RF-to-DC converter converts it to a voltage

(Vstor) across the external capacitor Cstorage which will bias the internal circuitry of the RF powered receiver. Once the bias voltage is high enough for the internal circuitry of the IC to operate, the ASK/FSK demodulator and digital decoder will check for the address of the device, and only if this is correct will the LDO be enabled (the EN signal goes high) and the Vout voltage rise up to bias the rest of the system (IoT node). Only at this point will the battery supply power. The EN signal is latched and will stay high as long as the equipment needs to operate. Once the apparatus has finished its job, the logic section can be reset through the "shdnb" pin; following this event the EN signal goes low and the power path between the battery and the output of the LDO takes on a very high impedance, implying negligible leakage current consumption (few pA) to ensure zero-energy "stand-by" operation or a fully turned-off device.

7 — **Quasi Nulling Stand-by in battery powered appliances**

Since the RF to DC transducer has a sensitivity of -18.8 dBm this is limiting the distance from which to turn on the remotely controlled appliance up to 8 meters when at 900MHz is transmitted a power of

0.5 Watt. In some cases this performance could be limiting and could he helpful considering a tradeoff solution in which the standby power consumption is not truly nulled but quite decreased. To do that

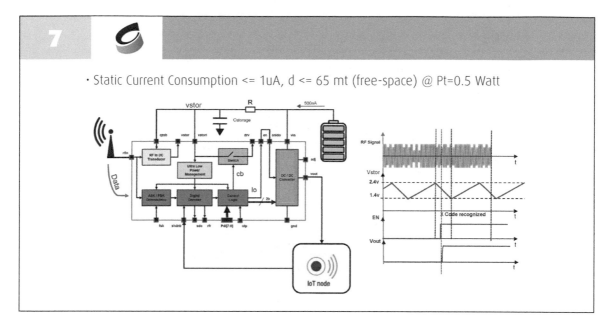

- Static Current Consumption <= 1uA, d <= 65 mt (free-space) @ Pt=0.5 Watt

the in the System on Chip we take advantage of the advanced performances of the ultra low power management that in the charging phase has a current consumption as low as 75 nA that allows to charge the capacitor Cstorage with a current, constantly provided by the battery, as low as 500nA. In this case the RF to DC transducer is no longer used and the sensitivity of the radio system is only limited by the ASK/FSK demodulator which is of -38 dBm. This higher sensitivity allows to extend the distance up to 65 meters in free-space with the same transmitted power of 0.5 watt.

8 Nulling Stand-by in battery powered appliances

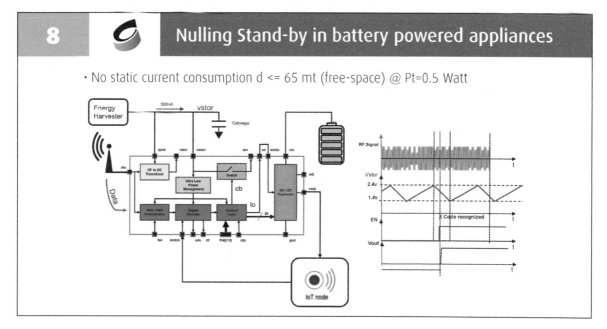

- No static current consumption d <= 65 mt (free-space) @ Pt=0.5 Watt

In some cases the 500nA of static current consumption could be provided by a different energy harvester that then internal integrated RF-to-DC converter. In this case the static current consumption of the remotely controlled device is nulled and the system will work as long as the Energy Harvester is able to provide energy with a higher performance in terms of distance and no static current consumption from the battery.

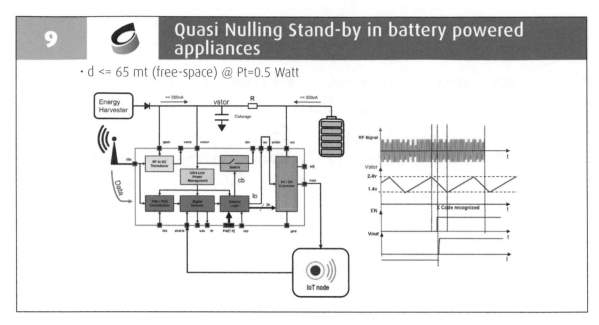

This circuit topology ensures that the system is always able to operate even if the energy harvester is not able to provide energy. If the energy harvester is able to provide energy with a current higher than 500 nA then the harvester will superimpose its current to bias the system and the battery will not source any current keeping no static current consumption while in standby. If the harvester cannot provide more current than the battery, this will source the current to keep the system in standby.

10   Nulling Stand-By in Europe would imply:

- **Saving 43 TWh / Year 4 Nuclear Power plants of 1.6 Gwatt**

- **19 Mtons of CO$_2$ / Year**

- **10 Million Euro / Year**

- **11% of Domestic power consumption**

To null standby power consumption in AC powered appliances is of grate benefit to increase the overall energy efficiency of the electrical appliances. In fact, it has been estimated that only in Europe the yearly energy consumption is about 43TWh, which is the equivalent of the energy generated by four, new generation, nuclear power stations of 1.6 Gwatt in a year. (www.eerg.it).

11 Nulling Stand-by in AC powered appliances

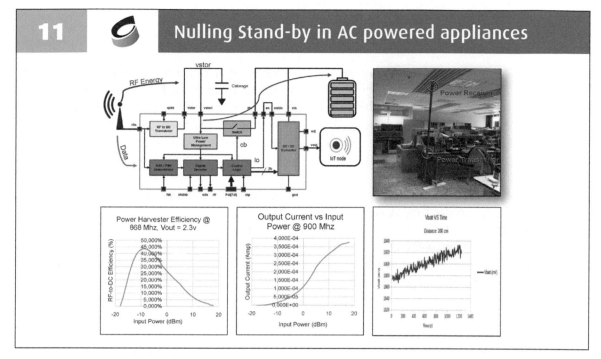

A charge and burst technique is used to turn on the high side transistor T1. When the system is off, though still plugged in, it does not consume any power and can be turned on remotely by using the energy from the RF signal. It is relevant to note that in this architecture, the auxiliary low power supply, and more in general, all the circuitry necessary to keep the system in standby, is no longer needed, with the advantage of reducing the cost of the solution and circuit complexity.

As shown in the Figure on the right the power of the RF signal charges the Cstorage capacitor. As soon as the voltage Vstor reaches 2.4 volts, the ASK/FSK demodulator and digital decoder extract the code, which if correct enables to bias high the internal "cb" signal and turns on the internal switch placed between the vstor and drv pins. As a consequence, the voltage between the base and the emitter of the transistor T1 is enough to turn it on. While transistor T1 is on, the capacitor Cvcc is charged until the startup threshold of the PWM controller is reached. At this point, the PWM controller is on and the entire power supply is self-biased through the dedicated winding (Lw) of the transformer.

12 Over the distance Wireless Battery Charger

Although node lifetime can be extended by minimizing the power consumption of electronics devices, by ensuring accurate design, using low or ultra-low power electronics components and by nulling, where possible, the standby power consumption at some point maintenance will be needed.

In particular, RF energy with wireless power transfer can be used to either directly power ultra-low power battery-free devices or to charge the batteries of small energy battery-powered devices. This can help to prevent or greatly reduce the cost of labour decreasing future maintenance efforts and costs to replace the batteries, and ultimately allowing ubiquitous IoT.

While the main advantage of RF energy is that is very pervasive allowing energy transferring with out of sight sensing devices, the main drawback is that the efficiency of the power transfer is very low, and for this reason is restricted to small energy devices. In fact, in a typical power transmission in the range of 900 MHZ in free space, the received power is about -30 dB (1/1000) the transmitted power after only 1 meter of distance and decays with -20 dB every 10 meters. Although far than efficient, RF energy transfer can be considered a very convenient way to provide power to small energy devices such as IoTwireless sensors.

The amount of radio frequency energy that can be stored in the battery to charge depends on many parameters such as the transmitted power, the gain of the transmitting antenna, the gain of the receiving antenna, the frequency of the transmitted RF signal, the distance between the RF transmitter and the efficiency of the RF Harvester.

In free space, the RF power received at the antenna can be calculated by using the equation for Friis transmission, as follows:

$$P\_R= P\_ (G\_T\ G\_R\ \lambda^2)/[(4\pi d)]^2$$

where P_R is the received power, P_T is the transmitted power, G_T the transmitting antenna gain, G_R the receiver antenna gain, λ the wavelength of the wave used, and d the distance between receiver and transmitter antenna.The power delivered to the battery is given by the product of the received power (P_R) at the antenna and the efficiency of the **RF** harvester (**Eff**).

$$P\_bat= Eff*P\_R$$

Considering a distance between transmitter and receiver of 2 meters, free space condition, a transmitted power of 0.5 W (27 dBm) at a frequency of 900 MHz and the gain of 1 for both antennas (**Gt=Gr=1**) the received power is $P\_R = 88\ \mu W$ (~ -10.5 dBm). Since the efficiency of the **RF** to **DC** transducer is ~ 40% when P_R = -10.5 dBm, the power delivered to the battery (**P_bat**) is 35 µW with an 20 uA average current delivered to a 2.3v battery.

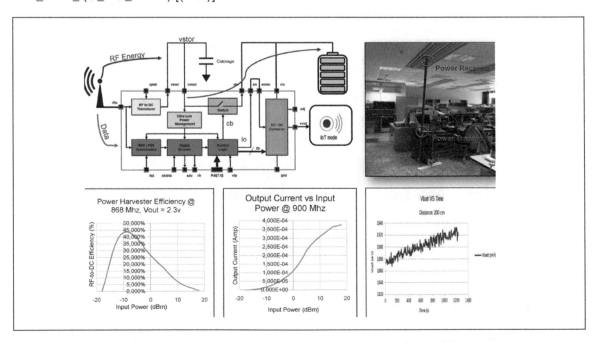

13 Powering Battery-Free Systems with WPT

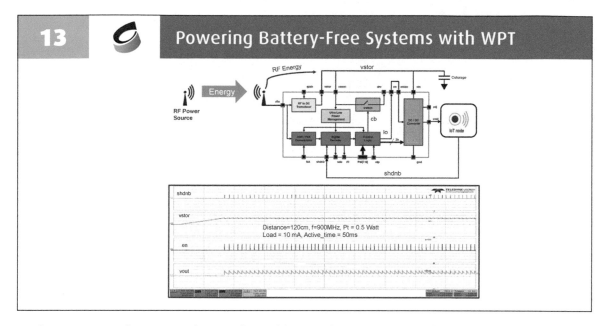

A battery powered sensor node is a disposable item which use is strictly limited by the life span of the battery. The main consequence of this is a high maintenance cost of the Wireless Sensor Networks especially when, as usually happens, the sensors are placed out of reach or hazardous places. The ultimate solution would be, when and where possible, to free the sensors from the batteries with the intent to eliminate the limitation of system lifetime due to the presence of the battery. In order to achieve this it is needed a specifically designed power architecture for wireless sensor node with an ultra power management able to handle currents in the nA order of magnitude. The energy could provided by several source such as RF, thermal, photovoltaic, vibrational, etc.

RF wireless power transfer is interesting due to the pervasiveness of RF energy, that can reach out of sight places even if must be considered that the efficiency of the transfer is very poor and this is the main reason why this technology is restricted primarily to small energy devices. In fact, as a consequence of the Friis transmission equation, in typical power transmission in the range of 900 MHZ in free space, the received power is about -30 dB (1/1000) the transmitted power after only 1 meter of distance and decays -20 dB every 10 meters, this is the main reason why this technology is mostly suggested in short range applications.

With reference to the figure in the bottom, the received Radio frequency energy is converted by the RF-to-DC energy transducer into a current that charges up the capacitor Cstorage causing the build-up of the voltage "vstor". When the voltage "vstor" reaches the maximum voltage of 2.4v, all the other internal circuits of the System on Chip turn on and the signal "enldo" goes high switcing on the Low Drop Output voltage regulator which provides a regulated voltage "vout" of 1.8v to the other units of the sensor node. At this point the system is completely on with a higher consumption which causes the voltage "vstor" to drop because of the imbalance between the harvested and required energy. In fact, in the typical case the Cstorage capacitor is required to source a current to the load in the order of milli amperes much higher than the harvested current which is instead in the order of micro amperes. As soon as the wireless sensor node has finished its activity the sensor unit will shut all the circuitry of the power unit down through the signal "shdnb". In this case the LDO is turned off and the voltage "vout" will go down causing the whole system to shut-down and the capacitor Cstorage goes back to be charged up through the harvested current.

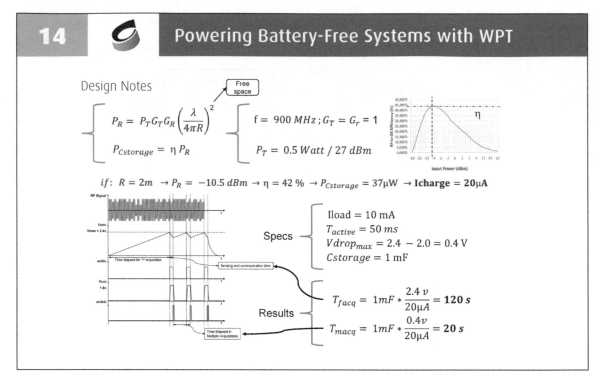

The value of the charging current, converted by the RF-to-DC from the radio frequency energy, depends primarily on the received power at the RF-to-DC which, in free space can be derived, as follow, by the Friistransmission equation:

$$P\_R= P\_\ (G\_T\ G\_R\ \lambda^2)/[(4\pi d)]^2$$

where **P_R** is the received power, **P_T** is the transmitted power, **G_T** the transmitting antenna gain, **G_R** the receiver antenna gain, λ the wavelength of the wave used, and **d** the distance between receiver and transmitter antenna. Thus, the power delivered to the capacitor Cstorage is given by the product of the received power (**P_R**) at the antenna and the efficiency of the RF harvester (**Ef**).

$$P\_Cstorage= Eff*P\_R$$

Considering a distance between transmitter and receiver of 2 meters, free space condition, a transmitted power of 0.5 W (27 dBm) at a frequency of 900 MHz and the gain of 1 for both antennas (Gt=Gr=1) the received power is **P_R = 88 µW** (~ -10.5 dBm). The efficiency of the **RF** harvester ~ 42% when **P_R** = -10.5 dBm, the power delivered

to the Cstorage capacitor (**P_Cstorage**) is 37 µW. In Steady state the average voltage at the Cstorage "vstor" is about 2 volts so the charging current is **Icharge = Pcstorage / vstor ≈ 20uA**. Regarding the load current we targeted a sensor unit able to carry out its sensing and communication activity with an average current of 50mA in a time window of 50msec with a maximum "vstor" voltage drop of 0.4 volts. This choice is to have enough voltage headroom to supply low power electronic devices requiring a typical bias voltage of 1.8v. Those requirements can be achieved by means of a low cost 1mFarad Cstorage capacitor. Just as an example of a feasible battery-free sensor node, assuming free space transmission, the RF power transmitter 2 meters far from the sensor node, which means a charging current of about 20uA, and the Cstorage capacitor completely discharged, once the RF power transmitter is turned on the sensor unit is able to perform its first acquisition after 120sec. All the acquisitions after the first one, as long as the RF power transmitter stays on, will be performed in a much quicker time since the Cstorage capacitor will keep a charge floor and will not be any longer completely discharged. Since the max allowed voltage drop is 0.4 volt, the time needed to get all the following acquisitions is 20 sec in the really worst case.

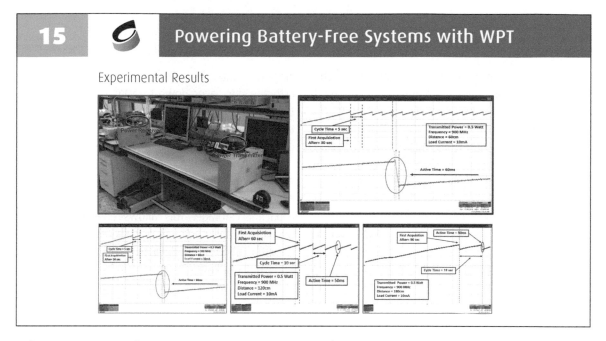

The setup consists of two units, a generic power transmitter board and the sensor node consisting of the System on Chip evaluation board as power unit, a sensor unit adopting a low power Bluetooth system as communication unit based on the STMicroelectronics IC BlueNRG-1 and a set of sensors such as accelerometer pressure and temperature sensor.

The power Transmitter is programmed to deliver 27 dBm (0.5 W) at 900Mhz. The pictures at the bottom of the slide show the voltage delivered to bias the battery-free sensor unit with highlight on the first acquisition time and cycle time at three different distances, 60cm, 120cm and 180cm respectively.

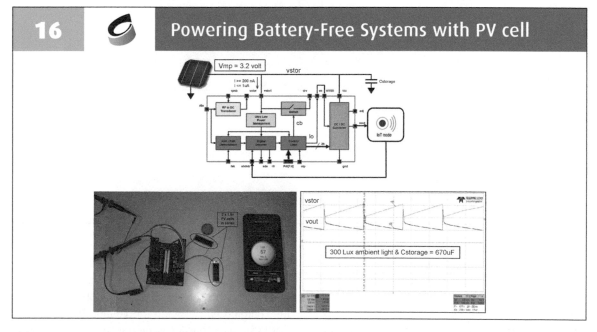

The Ultra Low power Management smartly manages the harvested energy, provided by the photovoltaic

cell, and delivers a regulated power supply to the internal circuitry of the system on chip as soon as

16

the amount of energy is enough to guarantee the active phase of the system. The power management consists of a nano-power circuitry that performs a voltage sensing on the storage capacitor to control the charging process and a micro-power circuitry that mainly provides the regulated power supply to the internal circuitry of the system on chip. During the charging phase when the voltage "vstor" across the Cstorage capacitor is below the programmable threshold Vmax (2.4v or 3.2v), the current consumption is as low as 75nA. Once the voltage "vstor" achieves Vmax the current consumption of the power management unit increases to micro-

amperes and the voltage across the capacitor Cstorage decreases down to a Vmin programmable threshold. As soon as the vstor voltage achieves the threshold vmin the micro-power circuitry is turned off thus enabling a new charging phase.

With such a low current consumption (75 nA) in the charging phase, a photovoltaic cell with a current capability of just 1μA can easily charge the storage capacitor Cstorage. This allows energy harvesting by using small photovoltaic cells, essential for the miniaturization of IoT devices, in environments with poor ambient light.

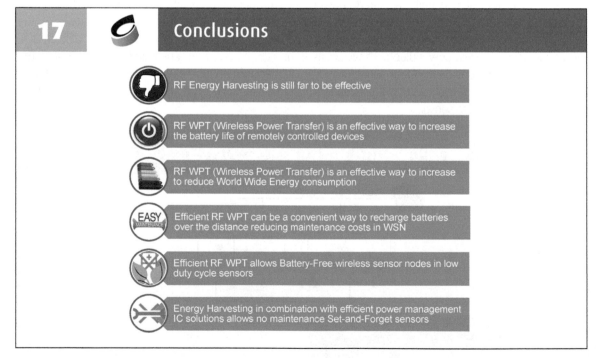

17 **Conclusions**

- RF Energy Harvesting is still far to be effective
- RF WPT (Wireless Power Transfer) is an effective way to increase the battery life of remotely controlled devices
- RF WPT (Wireless Power Transfer) is an effective way to increase to reduce World Wide Energy consumption
- EASY MAINTENANCE Efficient RF WPT can be a convenient way to recharge batteries over the distance reducing maintenance costs in WSN
- Efficient RF WPT allows Battery-Free wireless sensor nodes in low duty cycle sensors
- Energy Harvesting in combination with efficient power management IC solutions allows no maintenance Set-and-Forget sensors

The Ultra Low power Management smartly manages the harvested energy, provided by the photovoltaic cell, and delivers a regulated power supply to the internal circuitry of the system on chip as soon as the amount of energy is enough to guarantee the active phase of the system. The power management consists of a nano-power circuitry that performs a

voltage sensing on the storage capacitor to control the charging process and a micro-power circuitry that mainly provides the regulated power supply to the internal circuitry of the system on chip. During the charging phase when the voltage "vstor" across the Cstorage capacitor is below the programmable threshold Vmax (2.4v or 3.2v), the current

17

consumption is as low as 75nA. Once the voltage "vstor" achieves Vmax the current consumption of the power management unit increases to micro-amperes and the voltage across the capacitor Cstorage decreases down to a Vmin programmable threshold. As soon as the vstor voltage achieves the threshold vmin the micro-power circuitry is turned off thus enabling a new charging phase.

With such a low current consumption (75 nA) in the charging phase, a photovoltaic cell with a current capability of just 1µA can easily charge the storage capacitor Cstorage. This allows energy harvesting by using small photovoltaic cells, essential for the miniaturization of IoT devices, in environments with poor ambient light.

CHAPTER 04

Measuring and Analyzing Dynamic Current Profiles in Low Power Applications

Dr. Christoph Zysset

Application Manager, Computer Controls

The talk entitled *Measuring and Analyzing Dynamic Current Profiles in Low Power Applications* is motivated by the continuous development of Internet Of things infrastructure and applications or its flavours .

Lots of low power application are duty-cycled in order to save battery life. 1st order estimation is often misleading while building a whole sensor node the transient current waveform behaviour matter.

In this talk measurements techniques for power trace evaluation are discussed.

1 Low Power Applications

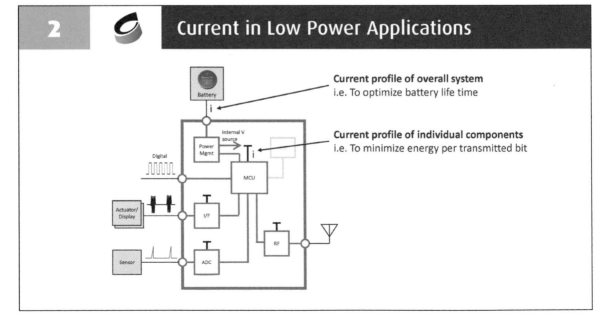

Many low power applications are closely related to the realm of the Internet of Things (IoT).

If we want or not, the IoT or its flavours affect us already and will affect us even more in the future.

According to the experts in this field, almost all areas of our daily life will be affected by the IoT.

One of the most important user expectations is a long life time of such a device after deployment.

Of course, life time has different aspects. In this presentation, we will focus on the aspect of how long a device is operational until it runs out of energy.

Just imagine, would you be happy if your would need to recharge your smart phone every two hours? Or your fit bit bracelet?

2 Current in Low Power Applications

Why are we or should we be interested in the current profiles of low power applications?

- The current consumption of a whole system (this might be for example a sensor node for an IoT application) helps to estimate the expected battery life and gives insights on how it could be optimized.

- The current profile of individual sub components can help to identify the main energy consumers and provides important information on how much energy is used to transmit for instance a bit, a byte or a measurement value of a sensor node.

3

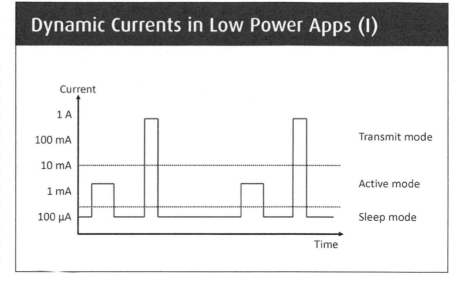

Dynamic Currents in Low Power Apps (I)

When we look a bit closer, we pretty quickly figure out that the currents we want to measure and to analyze in such a situation are by far not a DC current but rather something governed by the different states of the device or the components in this device.

Over time, the devices change the state they are in.

To save energy, they are put into a sleep mode when not in use.

Then they transit into an active state where for example a sensor value is obtained and processed.

Finally, there is a transmit mode.

This change of states or modes over time leads to three challenges.

4

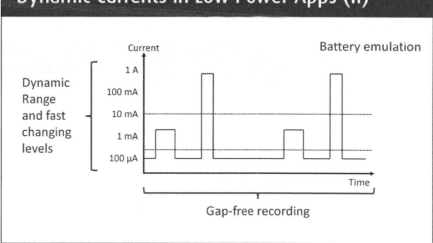

Dynamic Currents in Low Power Apps (II)

The first and probably most demanding challenge is the dynamic range. As we can see on the picture and you might already have experienced this in your applications, the current we want to measure can easily change over several decades.

The second challenge is the gap free recording over a longer period of time.

The last challenge is what I termed here «Battery emulation». Unfortunately, this is often not taken into consideration, but is actually important. The reason is that a battery voltage behaves differently from the output voltage of a lab power supply which is often used to power such devices during development.

In the next few slides, I would like to quickly discuss the traditional or popular measurement approach for these kind of measurement and would also like to point out several pitfalls where we run into conflicts with the three challenges we face.

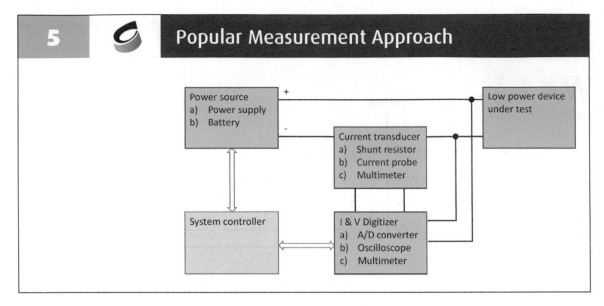

5   **Popular Measurement Approach**

A popular approach to measure the power consumption of a low power device or a low power component involves a power source, either in the form of a DC power supply or a battery.

The voltage applied to a device under test is measured using an ADC, a scope or a multimeter.

The current is measured using a shunt resistor, a current probe or a multimeter.

Often, a system controller like a PC or laptop is used to control the measurement setup and to store the acquired data.

Of course, this approach is attractive, because in most R&D labs multimeters and scopes are available and obtaining a shunt is easy. But, there are several points where you are in direct conflict with a dyanmic current measurement over a longer period of time.

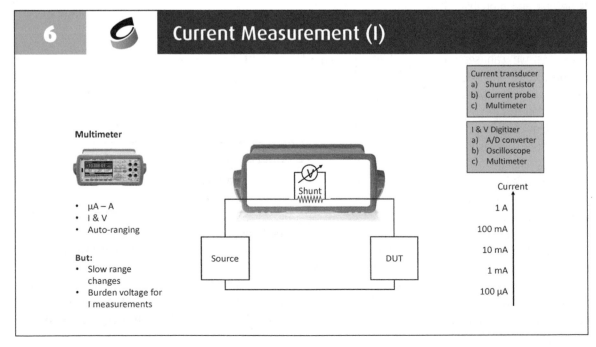

6   **Current Measurement (I)**

U sing a multimeter for measuring current is certainly not a bad choice. You can cope with the

aspect of a dynamic current, because multimeters are capable to measure from micro Amps up to Amps.

They achieve this by something called autoranging. This means that when one measures for instance a 3 micro Amp current, the multimeter automatically sets the measurement range to 10 micro Amps to achieve the highest resolution possible. If the current then changes to 8 mili Amps, the multimeter has to change its measurement range to 10 mili Amp.

But exactly this range change brings you into trouble. It usually takes up to milliseconds and if the device is already optimized to keep for example an active mode as short as possible, the multimeter is not capable to follow the transition which basically results in a very small bandwidth usually in the single

digit kHz range. This will strongly impair the current measurements during the transition from one state to another state.

Another potential disadvantage is the internal shunt resistor which is used to measure the current. The shunt resistor leads to a so called burden voltage, which leads to a reduction of the applied voltage on your device under test. As an example, let us assume that we want to power our device with 1.5 Volts. The shunt resistor can lead to a burden voltage of for instance 0.1 V which then leads to an effective voltage applied to your device of 1.4 V.

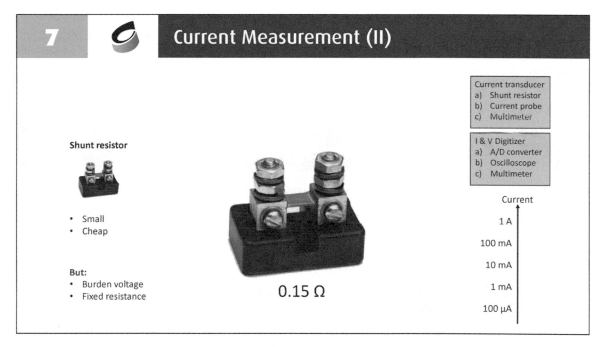

7 Current Measurement (II)

Shunt resistors are small and cheap and therefore an attractive choice.

But please be careful. A shunt resistor has always a fixed value. The chosen value is surely well suited for let's say a current in the 100 micro amp range. But when the current changes into the 100 mili amp

range, it is way too large and causes a tremendous voltage drop. This is of course not desired, because too large voltage drops will lead to a lot of other problems, like resetting of components or drop outs of the LDO, etc.

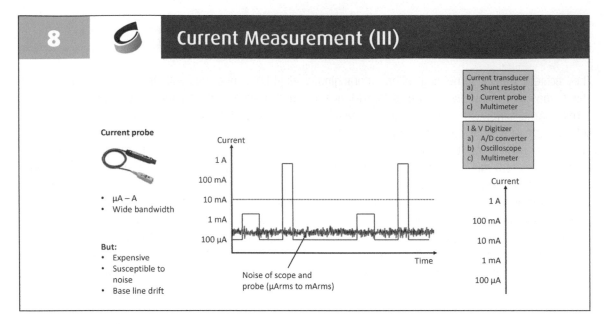

8 Current Measurement (III)

Current transducer
a) Shunt resistor
b) Current probe
c) Multimeter

I & V Digitizer
a) A/D converter
b) Oscilloscope
c) Multimeter

Current probe

- µA – A
- Wide bandwidth

But:
- Expensive
- Susceptible to noise
- Base line drift

Current

1 A
100 mA
10 mA
1 mA
100 µA

Time

Noise of scope and
probe (µArms to mArms)

Current

1 A
100 mA
10 mA
1 mA
100 µA

Current probes for oscilloscopes are attractive as well and on a first thought also beneficial. They can often cover wide ranges of current and provide a lot of bandwidth.

However, an oscilloscope and a current probe also produce noise. The sum of these two noise sources is in the range of several hundred micro Amps RMS to a few mili Amps RMS.

Sleep currents in the range of a few micro Amps or even in the nano Amp range will be completely covered by the noise.

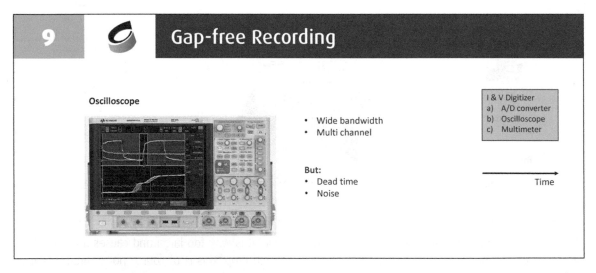

9 Gap-free Recording

Oscilloscope

- Wide bandwidth
- Multi channel

But:
- Dead time
- Noise

I & V Digitizer
a) A/D converter
b) Oscilloscope
c) Multimeter

Time

A straight forward choice for the digitizer is an oscilloscope. The reason is simple. It provides sufficient bandwidth, is compatible with the current probe and has multiple channels (which allows to sample current and voltage simultaneously).

However, gap-free recording is not possible with on oscilloscope. A digital oscilloscope, which is de facto the standard to day, has something called «dead time».

Dead time is the following.

10

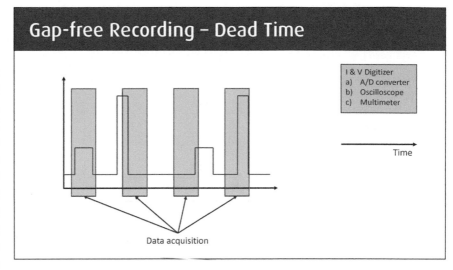

Gap-free Recording – Dead Time

I & V Digitizer
a) A/D converter
b) Oscilloscope
c) Multimeter

Time

Data acquisition

Basically, an oscilloscope does not sample all the time, although we might believe this because for our eye it looks like there is all the time new data on the screen. But in fact, an oscilloscope acquires a chunk of data and then pauses the acquisition for a certain time. In this time slot, called dead time, the scope processes the acquired data and re-arms its trigger system.

In essence, this means that we do not acquire a continuous data stream but rather segments out of it. This makes it very hard to for instance estimate the life time of the battery, because we get a quite incomplete picture of the current over time.

11

Battery Emulation

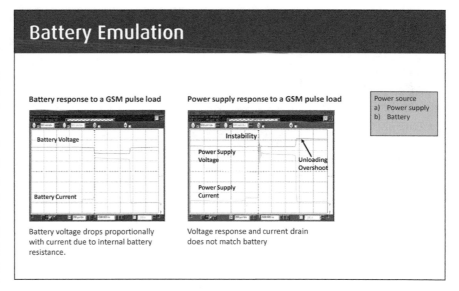

Battery response to a GSM pulse load

Battery Voltage

Battery Current

Battery voltage drops proportionally with current due to internal battery resistance.

Power supply response to a GSM pulse load

Instability

Power Supply Voltage

Unloading Overshoot

Power Supply Current

Voltage response and current drain does not match battery

Power source
a) Power supply
b) Battery

It makes a difference whether a device is powered by a DC power supply or a battery.

A DC power supply always regulates its output voltage to the set value, as can be seen in the right hand picture. An increase in the drawn current would result in a voltage drop, but the DC power supply immediately tries to counter act the voltage drop. Therefore, a ringing can occur. When the current returns back to its lower level, again the power supply tries to cope with the changed loading what usually results in a sharp increase of the voltage and then a slow return to the set value.

When the same device under test is powered by a battery, the voltage profile looks different. Due to the internal resistance of the battery, an increased current leads to a smaller output voltage. The change in the voltage levels occur almost immediately and with no or very little ringing.

Therefore, it makes often sense to use a battery emulator when testing a battery powered device to avoid that voltage drops caused by larger current consumption can lead to drop outs of the LDO or reset of components.

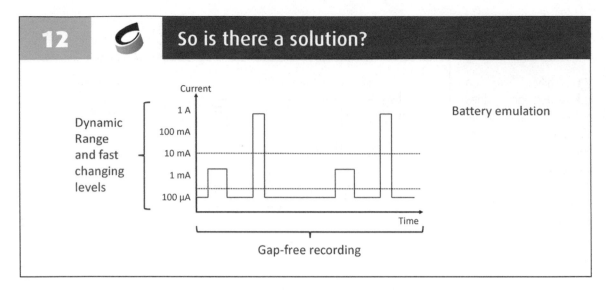

As we have seen in the previous slides, it is not easy to come up with a measurement setup which is able to cope with the dynamic range and the bandwidth of the current, which allows gap-free recording and has potentially the capability to emulate a battery.

So, the question is, is there a solution or a measurement system which is capable of doing all that?

The answer is «almost» or it depends.

Two approaches

There are two solutions, however, none of these solutions is a «one fits all».

If you are more interested in the current consumption and ultimately the power consumption of a whole system, the DC power analyzer can bring you closer to reality than a popular approach with DMMs, shunts, scopes and current probes.

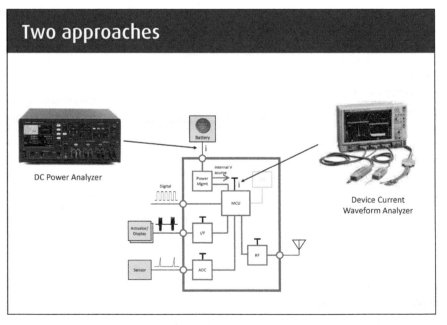

If you are more interested in the performance of a single component, i.e. a micro controller, a power amplifier or alike, then the Device Current Waveform Analyzer is a good choice.

However, as I mentioned, there is no «one fits all» approach and therefore, I would like to explain in the forth coming slides why these solutions are helpful for one task but not the other. In particular, why the power analyzer can help you to estimate battery lifetime and optimize the firmware of such a system and why the device current analyzer is a suitable tool to gain more insights into components.

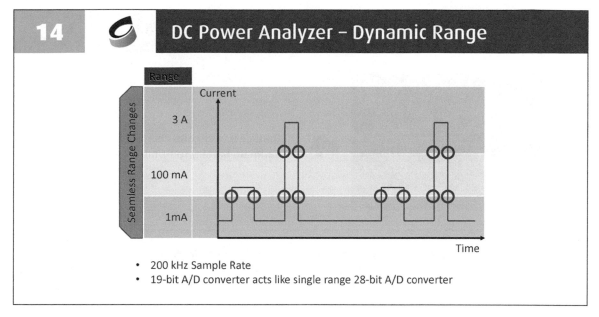

The seamless range change feature (patented technology of Keysight), allows to change the measurement range of the DC Power Analyzer between two consecutive sample points. Therefore, it is possible to characterize dynamic currents from the micro amp range up to the amp range with a bandwidth of 100 kHz.

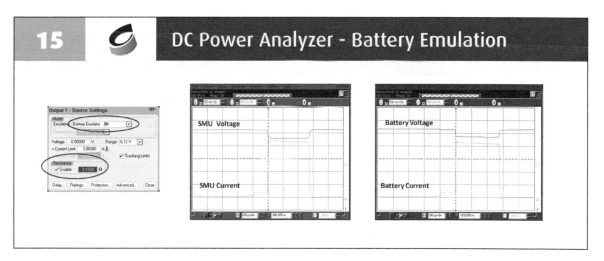

Since the DC Power Analyzer not only measures but also sources, the battery emulation feature is directly built into the unit. It can be used in a «battery emulator» mode where it is possible to set a value for the internal resistance.

Again, the GSM module which draws a current pulse is used. Comparing the current and voltage traces in both picture indicates that in the battery emulator the output of the DC Power Analyzer behaves like a battery.

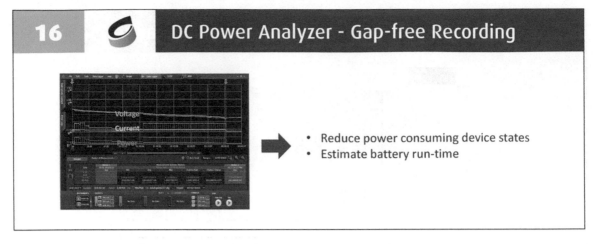

16   DC Power Analyzer - Gap-free Recording

- Reduce power consuming device states
- Estimate battery run-time

ogether with a ready-to-use piece of software, gap-free long term logging of voltage and current is possible. Long term logging can be done from seconds up to hours or even days.

This data can give you two insights into your system:

- The overall power consumption within the logging period can be used to estimate the

battery life time.

- Since no data decimation is performed when stored to a hard-drive, one can easily zoom into the current trace and match the current consumption over time to the different states the system goes through, i.e. The individual steps of the finite state machine in the micro controller.

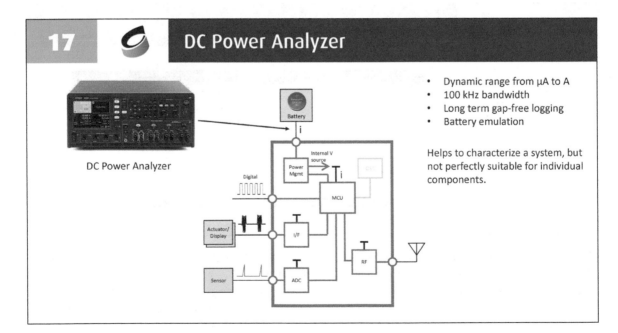

17   DC Power Analyzer

DC Power Analyzer

- Dynamic range from µA to A
- 100 kHz bandwidth
- Long term gap-free logging
- Battery emulation

Helps to characterize a system, but not perfectly suitable for individual components.

he DC Power Analyzer is certainly a wise choice when it comes to the characterization of systems. Not only because it can cope with the dynamics of currents in low power applications, but also because it is a single instrument and there is no need for

other equipment or software.

However, when it comes to characterizing an individual component like a micro controller, power amplifier or even a sensor element, usually the limited bandwidth is an impairment.

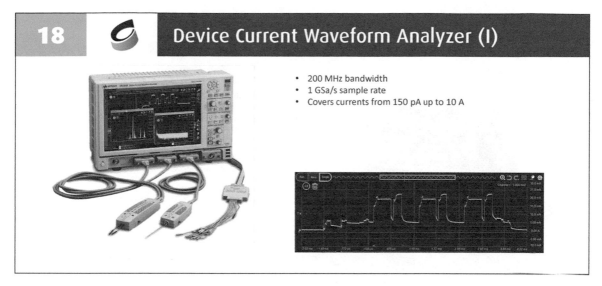

18 Device Current Waveform Analyzer (I)

- 200 MHz bandwidth
- 1 GSa/s sample rate
- Covers currents from 150 pA up to 10 A

Compared to the DC Power Analyzer, the device current analyzer offers more bandwidth and an even larger current range.

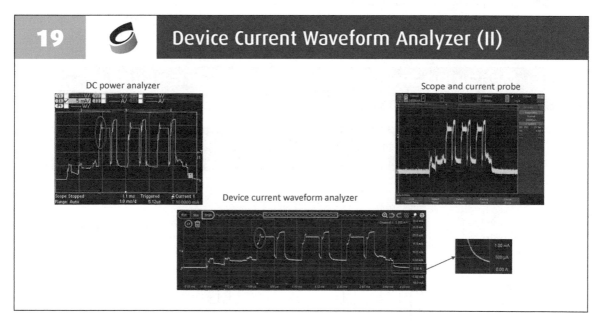

19 Device Current Waveform Analyzer (II)

DC power analyzer

Scope and current probe

Device current waveform analyzer

This slides shows the same waveform acquired with the DC Power Analyzer, a scope together with a current probe and the device current waveform analyzer.

Comparing the DC power analyzer waveform with the device current analyzer makes the advantage of more bandwidth visible: In the purple circle, the DC power analyzer measures some peaks which are actually not there (due to the longer integration time compared to the device current analyzer).

The difference between scope and device current analyzer is obviously the noise. From the scope data it would actually not be possible to estimate what the sleep mode current is, because the noise covers about half a division which results in 500 uAmp uncertainty. Using the zoom function (not software, it is a hardware zoom with a two channel probe) of the device current analyzer one can pretty accurately determinate the current in this region, whereas the scope data does not provide this information.

20   Device Current Waveform Analyzer (III)

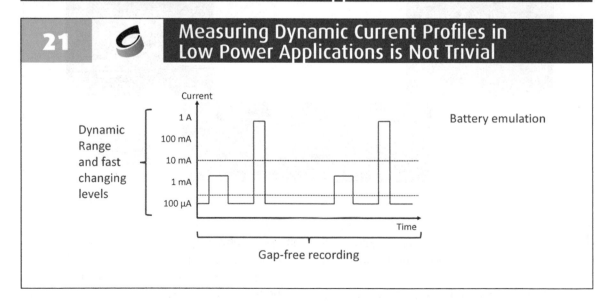

- 200 MHz bandwidth
- 1 GSa/s sample rate
- Covers currents from 150 pA up to 10 A

Suited to accurately measure the current consumption of individual components. However, it does not source power and does not support long term logging.

The device current waveform analyzer helps a lot to obtain current profiles of individual components like micro controllers, amplifiers or even to resistive RAM cells.

But one has to keep in mind, that this measurement system does not support long term logging out of the box and does not source power. An external power source for the device under test is needed.

Measuring and Analyzing Dynamic Current Profiles in Low Power Applications

21   Measuring Dynamic Current Profiles in Low Power Applications is Not Trivial

Take home message 1: Accurately measure such a current profile is not trivial. Coping with the dynamic range, fast changing current levels, gap-free recording and potentially battery emulation requires a quite sophisticated measurement systems and one has to be aware of its limitations.

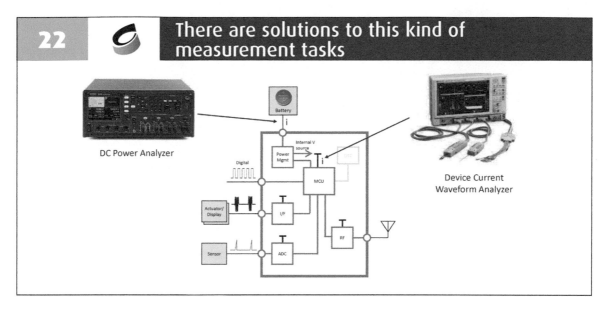

22 There are solutions to this kind of measurement tasks

DC Power Analyzer

Device Current
Waveform Analyzer

Take home message 2: There are two solutions to facilitate these measurement tasks.

CHAPTER 05

Challenges and Approached to Variation-Aware Digital Low Power VLSI Design for IoT

Prof. Andreas Burg

Ecole Polytechnique Fédérale
de Lausanne (EPFL), Switzerland

Typical IoT devices record data in remote locations with various types of sensors and relay this information to the cloud, where it is stored, analyzed, and combined with the information of other sensors. Unfortunately, the transmission of large amounts of data is typically expensive, not only in terms of communication bandwidth, but also in terms of energy. Hence, smart IoT sensors must extract and send only the essence from the sensor observations. This extraction requires complex near-sensor processing with good energy efficiency, which motivates the research for energy-efficient VLSI design in low-power IoT sensor nodes, using modern sub-100nm process nodes.

1

Low Power Digital VLSI Design

- **Voltage Scaling for Energy Efficient Low Power Circuit Design**

- **Managing Variability in Low-Voltage Integrated Circuits with**

- **Adaptive Body Biasing**

- **Low-Power Memories for Operation at Scaled Voltages**

This chapter is a review of recent ideas for energy-efficient low-power VLSI design, focusing on logic and on ultra-low-power embedded memories. In particular, we first introduce and discuss the basics of power and energy consumption in digital integrated circuits and we review voltage-scaling as the pre-dominant design approach to achieve energy-efficient low-power integrated circuits. We then discuss how uncertainties such as process and temperature variations require conservative design margins that are costly in terms of area, speed, and energy and how these design margins can be reduced by using the body-biasing capability of some process technologies such as FDSOI. In the last part, we focus on the issue of realizing embedded SRAM memories, which often constitute the major source of leakage and are the first point-of-failure at scaled voltages.

2

Power Consumption Bottleneck

In the last two decades, we have witnessed a shift from voice-centric to data centric communication, which came along with a need for more and more complex processing in various types of battery powered devices. The lifetime of these devices is ultimately limited by the energy consumption of the corresponding circuits and by the capacity of their batteries. Unfortunately, since the end of the 1990s, battery capacity has increased only slowly, and hence energy-efficient circuit design is our only means to enable more complex processing and longer battery lifetime.

- **Mobile devices: energy-efficiency**

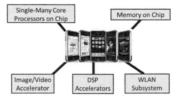

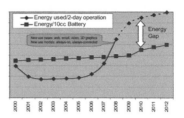

- **Battery capacity grows only very slowly**
  - **Boost in the 1990s due to Mobile Phone introduction**
  - **Capacity growth stalled since 2000 at the limit of Li-ion**
  - **Only 3%-7% annual improvement**

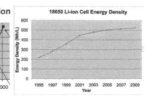

3 Power and Energy Consumption in CMOS

The energy consumption of CMOS integrated circuits in each clock period (or for each operation) is comprised of two main components:

On the one hand, leakage currents lead to static power consumption which decreases linearly with the supply voltage for the entire duration of the clock period. Hence,

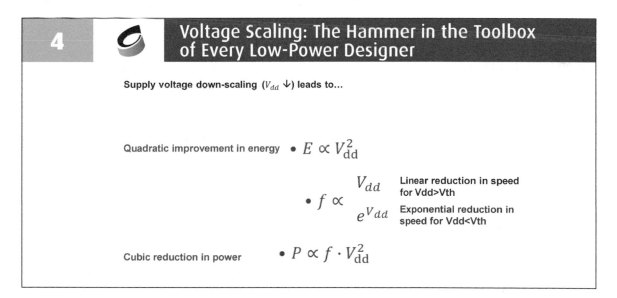

$$E = \left(\frac{C_L V_{dd}^2}{2} \beta + V_{dd} I_{leak} \frac{1}{f_{clk}} \right)$$

Total Energy Consumption

$V_{dd} I_{leak}$ **Static Power Consumption**

Dynamic Energy Consumption

Short Circuit Current

Switching $\frac{C_L V_{dd}^2}{2} \beta$

β : Activity factor
f_{clk} : Clock frequency

a shorter clock period and a reduced supply voltage are desirable to decrease the amount of leakage. On the other hand, the switching of the internal nodes drains a fixed amount of dynamic energy which is independent of clock period, but depends quadratically on the supply voltage.

4 Voltage Scaling: The Hammer in the Toolbox of Every Low-Power Designer

Supply voltage down-scaling (V_{dd} ↓) leads to...

Quadratic improvement in energy   • $E \propto V_{dd}^2$

• $f \propto$ V_{dd} Linear reduction in speed for Vdd>Vth

$e^{V_{dd}}$ Exponential reduction in speed for Vdd<Vth

Cubic reduction in power   • $P \propto f \cdot V_{dd}^2$

In particular the quadratic dependency of dynamic energy on the supply voltage (Vdd) motivates the use of a reduced Vdd to reduce the overall energy per operation and the energy consumption of integrated circuits. Unfortunately, decreasing the supply voltage, also comes with a reduction of maximum the frequency.

5

In the design phase, the loss in frequency at down-scaled voltages can often be compensated by a variety of architectural measures, such as pipelining or parallel processing, to still achieve a desired performance target. Yet, it is important to note that some of these design techniques (such as parallel processing) may also increase leakage or may exhibit a higher switching activity.

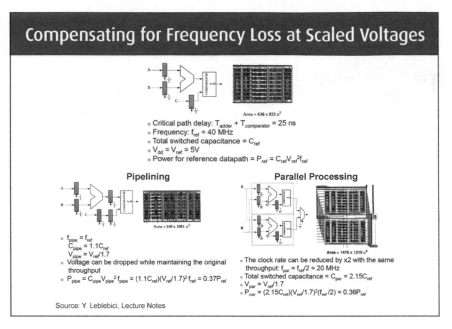

Compensating for Frequency Loss at Scaled Voltages

Area = 636 x 833 u²

- Critical path delay: $T_{adder} + T_{comparator} = 25$ ns
- Frequency: $f_{ref} = 40$ MHz
- Total switched capacitance = C_{ref}
- $V_{dd} = V_{ref} = 5$V
- Power for reference datapath = $P_{ref} = C_{ref}V_{ref}^2f_{ref}$

Pipelining

Area = 640 x 1081 u²

- $f_{pipe} = f_{ref}$
  $C_{pipe} = 1.1C_{ref}$
  $V_{pipe} = V_{ref}/1.7$
- Voltage can be dropped while maintaining the original throughput
- $P_{pipe} = C_{pipe}V_{pipe}^2f_{pipe} = (1.1C_{ref})(V_{ref}/1.7)^2f_{ref} = 0.37P_{ref}$

Parallel Processing

Area = 1476 x 1219 u²

- The clock rate can be reduced by x2 with the same throughput: $f_{par} = f_{ref}/2 = 20$ MHz
- Total switched capacitance = $C_{par} = 2.15C_{ref}$
- $V_{par} = V_{ref}/1.7$
- $P_{par} = (2.15C_{ref})(V_{ref}/1.7)^2(f_{ref}/2) = 0.36P_{ref}$

Source: Y. Leblebici, Lecture Notes

6

If ISO performance is not required and energy-efficiency is the primary design objective, voltage scaling offers an interesting design-knob, to trade energy-efficiency for processing speed. Starting from the nominal supply voltage of a process, reduced Vdd initially leads to quadratic gains in energy-efficiency with a linear decrease in throughput, i.e., a linear increase in the duration of each clock cycle or operation.

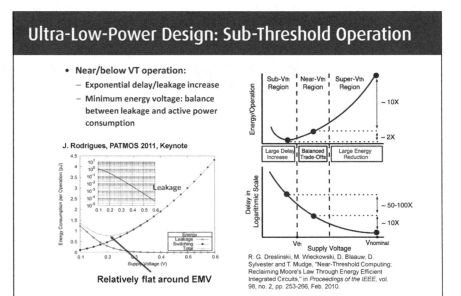

Ultra-Low-Power Design: Sub-Threshold Operation

- **Near/below VT operation:**
  - Exponential delay/leakage increase
  - Minimum energy voltage: balance between leakage and active power consumption

J. Rodrigues, PATMOS 2011, Keynote

Leakage

Energy
Leakage
Switching
Total

Relatively flat around EMV

Sub-Vth Region | Near-Vth Region | Super-Vth Region

~ 10X

~ 2X

Large Delay Increase | Balanced Trade-Offs | Large Energy Reduction

~ 50-100X

~ 10X

Vth   Supply Voltage   Vnominal

R. G. Dreslinski, M. Wieckowski, D. Blaauw, D. Sylvester and T. Mudge, "Near-Threshold Computing: Reclaiming Moore's Law Through Energy Efficient Integrated Circuits," in *Proceedings of the IEEE*, vol. 98, no. 2, pp. 253-266, Feb. 2010.

Unfortunately, as the supply voltage is further decreased down to or beyond the threshold voltage of the employed transistors, the delay of CMOS gates starts to increase rapidly, which requires significantly longer clock periods to meet timing constraints. At this point, the contribution of the leakage energy, which is the product of the leakage power with the rapidly increasing clock period, first starts to play a dominant role in the overall energy and eventually dominates over the diminishing active energy. A further reduction in the supply voltage beyond this point then leads to an overall increase in the energy per operation, which gives rise to the concept of a minimum energy point at the energy-minimum voltage of the circuit.

7

Leakage Power (I)

- **Transistors leak currents even when in off-state**
- **Sources for leakage**
  - Sub-threshold leakage
    - Dominant component in most circuits

  - Gate tunneling
    - Generally low, even in modern technologies due to high-k gate dielectrics
    - Decreases very rapidly with decreasing V_{dd}

  - Junction current
    - Generally low
    - Decreases very rapidly with decreasing V_{dd}

The important role of the leakage for low-power integrated circuits, which operate close to the energy minimum voltage, where leakage and active energy contribute equally to the overall energy budget motivates us to further study the sources of leakage power consumption and the search for design knobs that allow to reduce leakage.

In modern process technologies, below 100nm feature size, the dominant leakage component when a transistor is in its OFF state is the subthreshold leakage, while gate-tunneling and junction leakage generally play only a minor role.

8

Leakage Power (II)

- **Long channel deices (>130nm):** $I_{DS} = I_0 e^{\frac{V_{GS}-V_{th}}{v_t n}}$
  - I_{DS} mostly independent from Drain-Source Voltage
  - Leakage current depends strongly on $V_{GS} - V_{th}$
    - Decreasing threshold voltage increases leakage

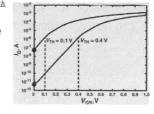

- **Impact of technology scaling on sub-threshold leakage (<130nm)**
  - Drain-Induced Barrier Lowering (DIBL): V_{DS} modulates threshold voltage
  - I_{DS} becomes a function of V_{DS}

$$I_{DS} = I_0 e^{\frac{V_{GS}-V_{th}+\lambda_{DS}V_{DS}}{v_t n}}$$

$$I_{leak} = I_0 e^{\frac{-V_{th}+\lambda_{DS}V_{DD}}{v_t n}} \longleftarrow \textbf{Voltage scaling reduces leakage}$$

When the transistor is OFF, i.e., when its gate-source voltage (Vgs) is zero, the subthreshold leakage of a MOS transistor depends exponentially on its threshold voltage Vth, which is one of the key process parameters.

A higher threshold voltage therefore generally reduces leakage and is therefore desirable to reduce energy, especially at low voltages.

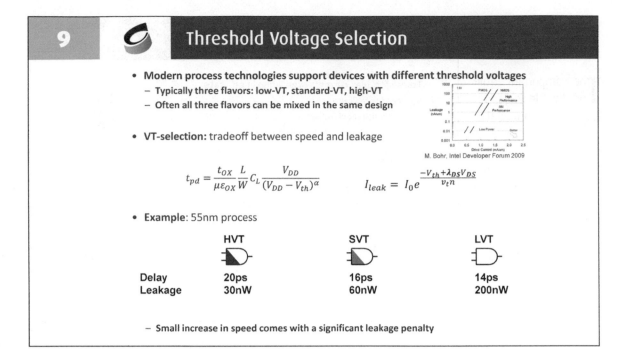

9   Threshold Voltage Selection

- **Modern process technologies support devices with different threshold voltages**
  - **Typically three flavors: low-VT, standard-VT, high-VT**
  - **Often all three flavors can be mixed in the same design**

- **VT-selection:** tradeoff between speed and leakage

M. Bohr, Intel Developer Forum 2009

$$t_{pd} = \frac{t_{ox}}{\mu \varepsilon_{ox}} \frac{L}{W} C_L \frac{V_{DD}}{(V_{DD} - V_{th})^\alpha} \qquad I_{leak} = I_0 e^{\frac{-V_{th} + \lambda_{DS} V_{DS}}{v_t n}}$$

- **Example:** 55nm process

	HVT	SVT	LVT
Delay	20ps	16ps	14ps
Leakage	30nW	60nW	200nW

  - **Small increase in speed comes with a significant leakage penalty**

Unfortunately, high-threshold devices not only reduce leakage, but also reduce the on-current, which reduces the speed of a circuit. When a circuit operates well above the threshold voltage, this impact on speed is significantly less than the impact on the leakage and therefore motivates the use of high-VT devices. However, for very low-voltages, high-VT devices can only provide sub-threshold on-currents, which again has a significant impact on their speed, which now decays also exponentially, which is counter productive.

Modern technologies generally provide a selection of devices with different threshold voltages. A careful selection of the right device type is therefore a key step in the design process to achieve the best possible tradeoff between leakage and active energy.

10   Variation Aware Design

Unfortunately, the selection between different device types is only possible at design time and is made based on assumptions of the device parameters. However, any variation in these design parameters (i.e., the threshold voltage) during manufacturing and operation may alter these parameters and thereby impact the balance between speed, leakage, and active energy.

11

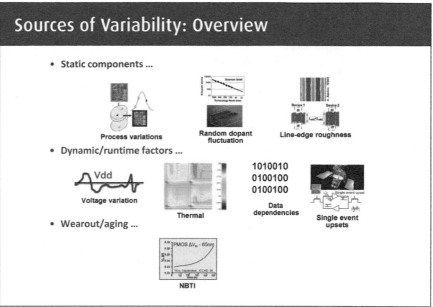

Uncertainties come from various sources and have different time scales. Variations in the manufacturing process have the longest time scale, since they lead to differences between individual manufactured circuits, but their impact remains constant for each manufactured circuit after production. Wear-out and aging of integrated circuits lead to slow variations in the circuit parameters over time-spans of years and decades. Finally, short term variations in the operating conditions, such as supply voltage and temperature even introduce uncertainties that vary during operation across a time span of milliseconds or even microseconds.

12

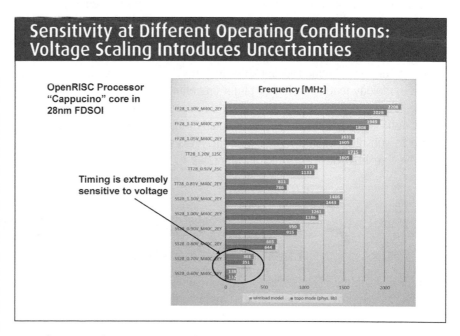

The impact if these variations is typically considered by analyzing a circuit under different operating conditions and with different device parameters that cover these uncertainties. The example shows how operating conditions and process parameters have a considerable impact on speed. A similar impact is observed when analyzing the leakage power consumption. In fact, we observe a correlation between leakage and speed, as slower speed is usually also associated with less leakage, while higher speed comes with larger leakage.

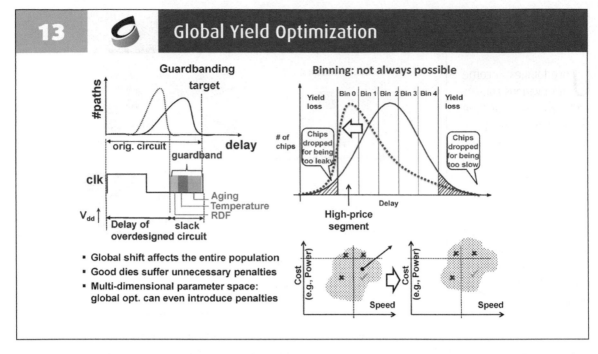

13   Global Yield Optimization

Guardbanding

- Global shift affects the entire population
- Good dies suffer unnecessary penalties
- Multi-dimensional parameter space: global opt. can even introduce penalties

Binning: not always possible

To ensure reliable operation across process parameter variations and under all environmental conditions, designers usually take a conservative approach, in which circuits are optimized for the worst case. In practice, this corresponds to adding worst case margins to timing requirements or designing a circuit for a higher operating speed so that even slow dies can meet timing requirements. In general, this methodology leads to a higher target operating voltage to enable this over design, which degrades energy efficiency. Alternatively, a designer may choose to substitute high-threshold voltage devices with low-threshold devices, which incorporates the timing margin that is necessary for slow devices, but leads to an excessive increase in leakage, especially for fast devices and in turn may violate leakage power or energy-efficiency design objectives.

However, we remind us that variations generally affect speed and leakage in static power consumption in opposite ways. An increase in delay (bad) typically comes with a lower leakage (good), while shorter delays (good) are associated with higher leakage (bad). Unfortunately, design-time measures are inadequate to exploit this correlation since the actual operating point is different for each manufactured circuit.

14   Adaptive Tuning: Basic Principle

Run-time adaptive circuits exploit the correlation between improved/degenerated timing and increased/reduced leakage. To this end, it is necessary to introduce mechanisms that allow to determine the timing and leakage parameters of each individual die after manufacturing and to adjust the leakage/performance tradeoff at runtime to restore the balance.

If this adjustment can be performed for each die individually, we avoid overcompensation of speed for fast dies at the cost of unacceptable leakage and enable the compensation of overly leaky, but fast dies by reducing their unnecessary timing margin. As a result, global margins on both leakage and timing

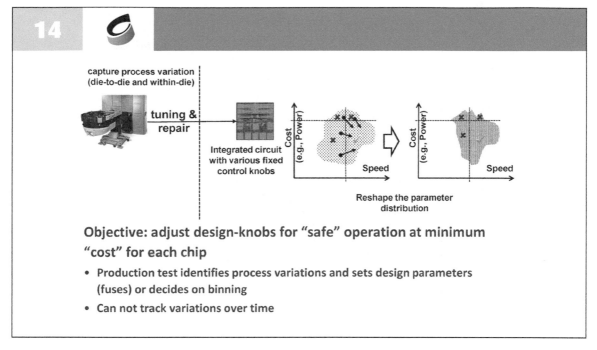

14

capture process variation (die-to-die and within-die)

tuning & repair

Integrated circuit with various fixed control knobs

Cost (e.g., Power)

Speed

Cost (e.g., Power)

Speed

Reshape the parameter distribution

Objective: adjust design-knobs for "safe" operation at minimum "cost" for each chip

- **Production test identifies process variations and sets design parameters (fuses) or decides on binning**
- **Can not track variations over time**

can be avoided which results on an overall lower power consumption and better energy efficiency.

As opposed to design-time measures, which can only move the cloud that represents the distribution of performance parameters in the scatter plot, this post-fabrication measure can re-shape the could, which leads to an overall better yield.

15

Body Bias Modulates Threshold Voltage

Runtime adaption of the tradeoff between leakage and the speed of a circuit can be enabled by modulating the threshold voltage of its transistors through their back-gate voltage, which is referred to as (adaptive) body-biasing. Under nominal conditions, the body bias of a transistor is zero in conventional CMOS logic. A forward body bias decreases the threshold voltage and thereby increases speed, but also leakage. A backward body

- **Body of the transistor is often connected to the source (no body bias)**
- **Introducing a body bias modulates threshold voltage**
  - **Forward Body Bias (FBB): increases threshold voltage**
  - **Reverse Body Bias (RBB): reduces threshold voltage**

$V_{BS} < 0$ $V_{BS} > 0$

FBB

RBB $V_{BS} > 0$

- $V_{th} = V_{th0} - \lambda_{BS} V_{BS}$

$V_{th0} < 0$ $V_{th0} > 0$

- **Application to logic**

Forward Body Bias   Reverse Body Bias

PMOS

PMOS

NMOS

NMOS

BB Gen

BB Gen

- **FBB improves speed, but also increases leakage**
- **RBB reduces leakage, but also increases delay**

bias decreases the threshold voltage which reduces leakage, but also decreases speed.

16 Body Bias for Leakage Reduction

To dynamically adjust the body voltage at runtime, the body voltages must be generated on the chip, but since only a change in the body voltage requires notable currents, this voltage generation typically comes with a low overhead. The ability to adjust body voltages at run-time, further allows to adjust the tradeoff

- **Dynamic body bias:**
  - Generation of bias voltages involves overhead
    - Voltage regulators (usually only very small currents)
    - Generation of negative or >VDD voltages (switching regulator)

- **Adjust BB (threshold voltage) dynamically for sleep-mode and active-mode**

- **BULK CMOS:**
  - Effect of body bias decreases for technologies below 100nm
  - FBB is limited to ~300mV to avoid operating junction diodes in forward direction

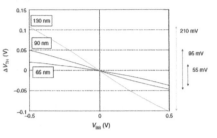

between leakage and speed to support different modes of operation, such as sleep modes, where clock frequencies are low.

Unfortunately, the impact of the body voltage on the threshold voltage depends on the process technology. In bulk-CMOS, more advanced technology nodes usually have a lower body factor, which limits the impact on body biasing to older process generations.

17 Body Bias in FD-SOI Technologies

- Thin un-doped channel
- Thin buried oxide layer (BOX) isolates channel from the body (back-gate)
- Body contact allows to control the back-gate

- **Advantages:**
  - Un-doped channel avoids V_{th} variation due to RDF compared to Bulk-CMOS

- **Threshold voltage modulation through back-gate**
  - Large back-gate voltage range and high V_{th} sensitivity: ~60 mV/V

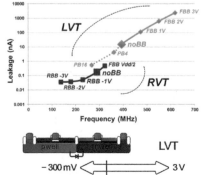

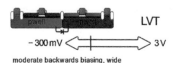

Source: P. Flatresse, ST Microelectronics

An alternative to bulk-CMOS is offered by FDSOI technology. This type of technology typically comes with strong body factors, even in combination with small feature size transistors. Hence, it provides a great platform for runtime compensation of variations.

18

Adaptive Tuning: Basic Principle (I)

With body-biasing providing a powerful knob for adjusting the tradeoff between leakage and delay of a circuit at run-time, the difficulty now lies in the sensing of the actual circuit performance parameters and the translation of the sensor output into a proper setting of the backgate transistor voltage. The most basic approach simply obtains circuit parameters after production during testing. However, this only allows to compensate for manufacturing variations, while variations due to aging, temperature, or voltage must be covered by worst-case margins.

capture process variation (die-to-die and within-die)

tuning & repair

runtime design-tuning knobs (Voltage/Frequency)

"Sensors"

Voltage

Temperature

Aging

Integrated circuit with various control knobs to adjust operation

Measure operating conditions

Controller

Determine appropriate measures LUT

Capture slow runtime variation

- Objective: adjust design-knobs for "safe" operation at minimum "cost" for each chip at any point in time
  - Initial testing provides
    - a baseline calibration for process variations
    - tuning parameters as a function of sensor readings => long and complex testing required to characterize many operating points (interpolation difficult)

By integrating sensors onto the chip that measure these variations at run-time, a controller can also compensate for these type of dynamic variations (open-loop).

19

Adaptive Tuning: Basic Principle (II)

The difficulty in this open-loop configuration lies in the translation of the sensor readings into an appropriate bias setting due to the large parameter space and a lack of full and accurate visibility of all parameters that have an influence on the relationship between the sensor readings and circuit performance.

tuning & repair

runtime design-tuning knobs

"Sensors"

Controller

Integrated circuit with various control knobs to adjust operation

Check for the impact of variability

Determine appropriate measures

capture process variation (die-to-die and within-die)

capture runtime variation

- Objective: adjust design-knobs for "safe" operation at minimum "cost" for each chip at any point in time
  - On-chip sensors check directly on the operating margin and detect errors

An elegant method to compensate for these unknowns is to directly, rather than indirectly sense the actual circuit performance parameters such as speed and leakage and to adjust the run-time tuning knobs (body bias voltages) with a close-loop feedback control.

20

A n example for this approach was published in a 2002 paper from Intel, where circuit timing sensors in the form of delay lines provide information on the circuit speed. This information is used to adjust the body voltage. Without this adjustment, different circuits provide a large spread in the leakage/frequency plane and many circuit realizations lie outside the specifications. As a result, the yield is low. With the feedback control enabled, circuits with a too low speed receive a forward body bias which compensates for the loss with a small, but

Electrical Knobs: Adaptive Body Bias

Maximize clock frequency under total power constraint

- V_{th} variation determines leakage/freq. operating point
- Adjust V_{th}: forward/reverse body bias (FBB/RBB)

RBB: leakage ↓, speed ↓   **FBB: leakage ↑, speed ↑**

Tschanz, James W., et al. "Adaptive body bias for reducing impacts of die-to-die and within-die parameter variations on microprocessor frequency and leakage." Solid-State Circuits, IEEE Journal of 37.11 (2002): 1396-1402.

acceptable leakage penalty. At the same time, leaky circuits receive a backward body bias which reduces leakage, but avoids unnecessary timing margins. The overall yield is improved.

21

T he simplest approach to sense the actual speed of a circuit through an on-chip sensor is the us of a canary circuit, which can be either a ring oscillator or a replica of the critical path of the circuit. The delay of this circuit is then measured through a time-to-digital converter. Unfortunately, the delay of different gates is affected slightly differently by parameter variations. Hence, a single canary circuit can often not fully reflect the delay scaling behaviour of all paths in a complex circuit.

Sensors: PVT Tracking Based on Delay Lines

Delay lines capture delay increase due to global PVT variations

- **Single instance *cannot* capture intra-die variations, local supply drops, crosstalk**

Delay variation of complex gates is significantly different from that of an inverter, especially at low voltages

Time-to-digital-converter to measure delay as input to closed-loop voltage control

Dhar, Sandeep, Dragan Maksimović, and Bruno Kranzen. "Closed-loop adaptive voltage scaling controller for standard-cell ASICs." *Proceedings of the 2002 international symposium on Low power electronics and design.* ACM, 2002.

It can further not capture variations that originate for example from different temperature changes in different corners of a large die.

22

To address these issues, more complex delay sensors integrate a variety of different canary circuits who's individual measurements can be combined to provide a better estimate of the worst case path in a complex design. Temperature variations across the die can be accounted for by combining multiple such sensors, as demonstrated in a 2007.

Sensors: Calibrated Critical Path Monitors [IBM Power6]

- *Distributed across the chip* to better capture local variations

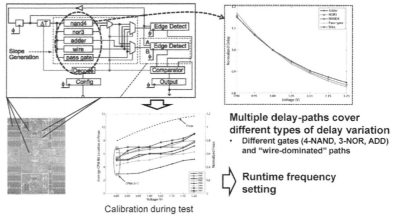

Multiple delay-paths cover different types of delay variation
- Different gates (4-NAND, 3-NOR, ADD) and "wire-dominated" paths

➡ **Runtime frequency setting**

Calibration during test

Drake, Alan, et al. "A distributed critical-path timing monitor for a 65nm high-performance microprocessor." *Solid-State Circuits Conference, 2007. ISSCC 2007. Digest of Technical Papers. IEEE International.* IEEE, 2007.

23

The use of replica circuits a simple approach for tracking timing variations, but the design of a replica that accurately follows the speed of a complex circuit is difficult. Error detection sequentials avoid the use of a separate canary circuit. Instead, the sequential elements in a synchronous circuit are extended to detect marginal timing

Sensors: Error Detection Sequentials (In-Situ Error Detection)

Basic idea: detect timing violations directly at each path endpoint to capture also local variations, cross-talk, and data dependent delay variations

- **Detect any change of data after the clock edge (late-arrivals)**

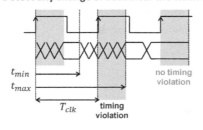

t_{max}: worst case path delay

t_{min}: contamination delay

$$t_{CLK} < t_{max}$$

$$t_W = t_{max} - t_{CLK}$$

Tradeoff between t_W and t_{min}

$$t_{min} > t_W + t_{hold}$$

- t_W ↑: more margin for variability, but more extra buffers to guarantee t_{min}

conditions or timing violations. Such violations indicate that a circuit is too slow and can trigger a corresponding action to increase the speed for example through an increase of the forward body bias.

24

Various examples of such sequentials with timing-error detection have been reported in the literature, based on an original idea from 2003, known as RAZOR.

Sensors: Error Detection Sequentials

Examples: 1st generation
- **Data stored with an edge-triggered FlipFlop**
- **A shadow latch remains open during the timing violation window**
- **Error signal generation by comparing data and shadow-latch output**

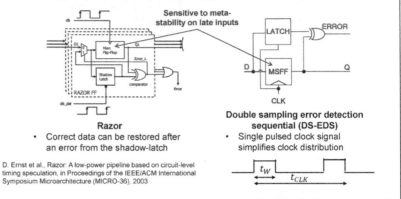

Razor
- Correct data can be restored after an error from the shadow-latch

Double sampling error detection sequential (DS-EDS)
- Single pulsed clock signal simplifies clock distribution

D. Ernst et al., Razor: A low-power pipeline based on circuit-level timing speculation, in Proceedings of the IEEE/ACM International Symposium Microarchitecture (MICRO-36), 2003

25

When comparing the use of replica circuits with the use of error detecting sequentials, we note that RAZOR type approaches generally allow to better sense the circuit timing behaviour and therefore allows to operate close to the limits and to avoid margins. However, this comes at the price of an increased overhead since at least any near-critical path of the circuit must be instrumented with error detection. Yet, an overall conclusion, which of the two approaches is more is difficult to obtain

Architecture Level Knobs: EDS vs. TRC based Detection

Compare Error Detection Sequentials vs. Timing Replica Circuits (TRC)

Example: Microprocessor [Intel]
- **EDS is more complicated to implement**
- **EDS involves more overhead than TRCs since sequential elements are more complex**
- **EDS reduces the number of replay events and provides tighter margin**
  - Captures data dependent delays
  - Does not require guard-bands to cover local variations

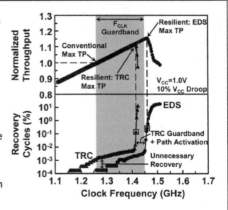

Bowman, Keith A., et al. "A 45 nm resilient microprocessor core for dynamic variation tolerance." *Solid-State Circuits, IEEE Journal of* 46.1 (2011): 194-208.

and generally also depends on the design under consideration and its desired operating point.

26   Low Power Memories

Most complex designs are composed of logic and memory. While logic is built from standard cells, memories are based on highly optimized macros to achieve a high density.

27   ITRS Roadmap Prediction on the Role of Memory

A study of a wide range of different designs show that memories often occupy a large percentage of the overall area. For future systems and technology generations, the ITRS roadmap predicts a further increase in the amount of memory compared to the amount of logic, which renders the design of low-leakage

- **Amount of memory (percentage) increases**
- **Leakage becomes increasingly relevant in modern technologies**

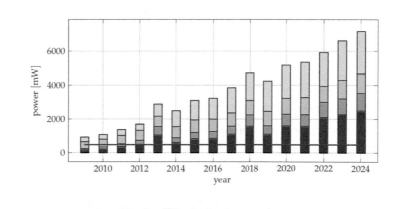

low-power memories one of the foremost concerns in low-power design.

28   SRAMs Consume a Considerable Amount of Power

Worse than the significant contribution to area is the fact that memories are typically reponsible for the majority of the active energy and leakage power consumption of IoT devices, which often spent the majority of their life-time in sleep mode.

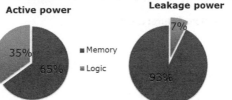

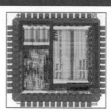

typical embedded processor

- **For embedded processors, memories occupy a large percentage of the silicon area**
- **Active mode:**
  - **Data and program memory can consume 2/3 of total power**
  - **Low-frequency: SRAM leakage becomes visible**
- **Sleep modes:**
  - **Generally, no power gating to retain SRAM content**
  - **SRAM leakage becomes dominates system power consumption (3-4 pJ/bit in 180nm): 32kByte -> 400nW @ 1.8V**

29

 SRAMs at low voltages

Furthermore, conventional SRAM memories contribute indirectly to high overall power consumption, as they limit the potential for voltage scaling since conventional 6-transistor SRAM cells are particularly sensitive to device variations. This turns them into the first point of failure when the supply voltage is reduced. For example, conventional SRAM memory macros are often only specified down to 0.6-0.7V, which lies far above the minimum supply voltage of conventional logic.

Standard 6T SRAM becomes unreliable at scaled voltages due to process variation
- **28nm technology: 6T SRAM functional down to 0.7V**

6T SRAM cells rely on rationed circuits
- **Mismatch between access and pull-up/down transistors during cell access**

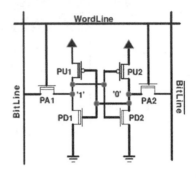

30

SRAM failure mechanisms at low voltages

The main reason for this limitations lies in the fact that 6-T SRAM bit-cells are ratioed cells which are sensitive to process parameter variations. Such variations reduce the design margins and lead to read-, write-, or retention-failures.

- **Write-failures: inability to write**
- **Access-failures: inability to read**
- **Read-failures: pre-charged bit-line flips cell content during read**
- **Hold-failure: content flips without access**

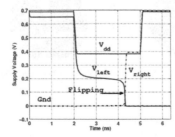

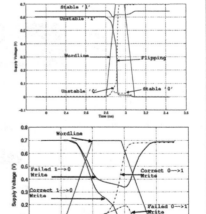

31

The amount of available margins is illustrated by the static noise margin of a bit cell, which provides an indication of the ability of this cell to reliably keep its content. As the supply voltage reduces, this margin shrinks, as well as the corresponding margins for reliably reading from or writing to the cell.

SRAm Stability Analysis at low Voltages: Static Noise Margin

- Static noise margin (SNM): maximum amount of voltage noise that can be introduced at the output f the back-coupled inverter pair while maintaining stability
  - Draw transfer characteristics of the two inverters
  - Diagonal of the largest rectangle that can be embedded between the two curves

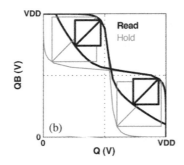

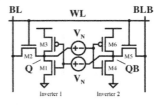

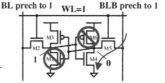

Calhoun, B.H.; Chandrakasan, A.P., "Static noise margin variation for sub-threshold SRAM in 65-nm CMOS," *Solid-State Circuits, IEEE Journal of* , vol.41, no.7, pp.1673,1679, July 2006

32

An effective idea to improve the reliability of an SRAM bit-cell is to revert to the less dense circuit structure of a latch, which avoids the ratioed design by breaking the feedback loop and decouples the storage node from the output during read through a dedicated driver. While this cell requires significantly more transistors, it has better noise margins and operates more reliably and remains operational even far below 600mV.

Latches: the perfect low-voltage SRAM cell

Most reliability issues of 6T SRAM are avoided by latch
- **Write failures: unusually strong keeper ↔ feedback disabled**
- **Read failures: degradation of SNM ↔ isolated output**
- **Hold failures: SRAM bitcell = latch in non-transparent phase with still good SNM at VDD=300mV**

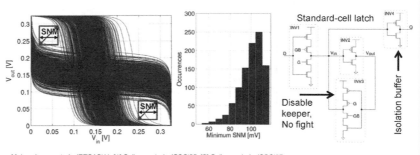

Meinerzhagen *et al.*, JETCAS'11; [1] Calhoun *et al.*, JSSC'05; [2] Calhoun *et al.*, JSSC'07

33

Custom Cell: Low-Leakage Latch with Tri-State Output

To reduce leakage further, latch based memory cells also be optimized specifically toward this objective, for example by stacking transistors or by stretching their length.

Best practice for low leakage

1. Lowest number of V_{DD}-ground paths
2. Highest resistance on each such path

➤ Tri-state buffers
➤ Stacking & stretching for inverters

Stacking factor: max 2

Channel length stretching: $2L_{min}$

Convert output buffer to tri-state buffer to avoid static CMOS muxes

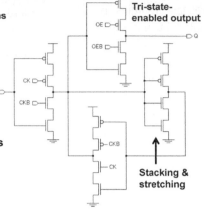

Tri-state-enabled output

Stacking & stretching

34

SCMs best practice

Write and read access in latch based memories are realized through clock gating of the memory latches and through logic or tri-state based multiplexers.

Write Logic
➤ *Clock-gates (b):* smaller and less power than *enable flip-flops (a)*

Read Logic
➤ Above-VT
  ✓ *Multiplexers (c):* smaller, faster, and less power than tri-state buffers
➤ Sub-VT
  ✓ *Tri-state buffers (d):* less leakage (energy) than multiplexers

Array of Storage Cells
➤ *Latch* arrays smaller than *flip-flop* arrays, but longer write-address setup time

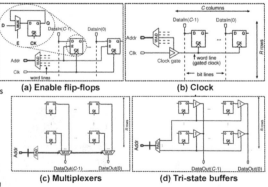

(a) Enable flip-flops

(b) Clock

(c) Multiplexers

(d) Tri-state buffers

The so constructed standard-cell based memory macros offer a wide range of advantages over 6T-SRAM based macros. The standard cells can also be placed either with the surrounding logic, or with guided placement, which exploits the regular structure of their design.

Standard-Cell based Memories (SCMs)

Assemble small memories from standard-cells
- **Robust against voltage scaling**
- **Inherently low power**

Advantages beyond robustness and power
- Generic description in any HDL
- Any desired size is possible
- Modifications at design time
- Portability (unless custom cells)
- Fine-granular organizations
- **Automatic or guided placement**
- Merge with logic (where appropriate)
- Avoid power routing

Main drawback
- **Area (if storage capacity > 1kb)**

Meinerzhagen *et al.*, MWSCAS'10
Roth, Meinerzhagen *et al.*, A-SSCC'10

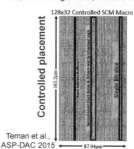

128x32 Controlled SCM Macro

Teman et al., ASP-DAC 2015

Overall, Standard-Cell Memories (SCMs) provide a significant advantage in terms of active and leakage power compared to 6T SRAM macros at the expense of a 2-3x larger area. While this area penalty is clearly prohibitive for large memories, it is often more than acceptable for small macros.

SCMs Outperform 6T SRAM Macros und nominal Voltages

- **Lower read- and write-energy**
- **Lower leakage power**
- **Sometimes even better access speed**

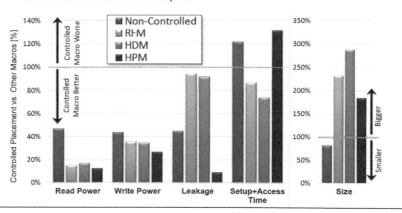

37

The figure shows the block diagram of a standard-cell based memory based on latches optimized for low-leakage with an integrated tri-state driver for multiplexing of the outputs.

Test Chip: Architecture of Low-Leakage 4kb SCM

➢ Write logic uses clock-gates

➢ 3-state buffers used for read operation are integrated in low-leakage latch

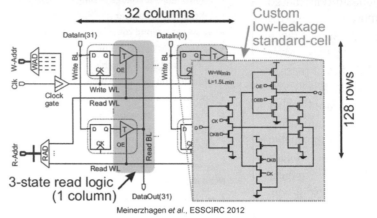

Meinerzhagen *et al.*, ESSCIRC 2012

38

The design was realized in a 180nm CMOS technology and tested across a temperature range from 27-37 degrees, which is a typical range for implanted ultra-low power devices.

4kb SCM Test Chip in LP-HVT 65nm CMOS

Chip microphotograph and zoomed-in layout picture

Area cost of 12.7 µm² per bit (including peripherals)

Scan-chain test interface

Functionality verification:
W/R random and checker-board patterns

Oven to control temperature:
27 or 37°C

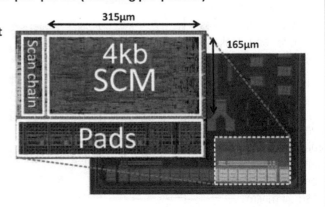

39

Sub-VT SCM: Insights and Leakage Breakdown

Simulations show the energy consumption in sub-threshold operation. The energy minimum point can be clearly identified. We note that different memory configurations place this energy minimum operating point at different voltages, which calls for a joint optimization of the memory with the surrounding logic to target a global minimum.

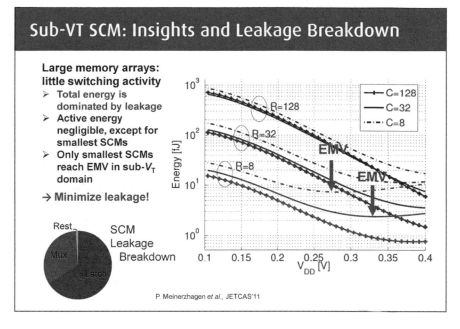

Large memory arrays: little switching activity
➢ Total energy is dominated by leakage
➢ Active energy negligible, except for smallest SCMs
➢ Only smallest SCMs reach EMV in sub-V_T domain

→ **Minimize leakage!**

SCM Leakage Breakdown

P. Meinerzhagen et al., JETCAS'11

40

Silicon Measurements: Active Energy is 14 fJ/bit-access

Similar results are also visible from the measurements of the manufactured array. At the energy minimum voltage of 500mV, the circuit operates at a maximum frequency of 110kHz. Larger supply voltages, lead to a rapid increase in circuit frequency, but active energy dominates. For smaller supply voltages, the operating frequency degrades further to only a few kHz and leakage becomes the dominant factor.

Measured energy per bit-access performed at maximum speed

Measured energy minimum is 14fJ/bit at 500mV, 110kHz

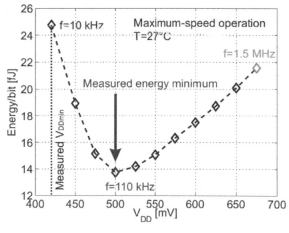

41   Silicon Measurements: Leakage Power is 500fW/bit

At VDDhold=220mV, data is correctly held with a leakage power of 425–500fW per bit (best and worst out of 4 measured dies)

At 37°C (typical for biomedical implants)

- VDDmin=400mV (instead of 420mV at 27°C)
- Maximum operating frequency doubles
- But: higher leakage power

- Low retention voltage is key for low power

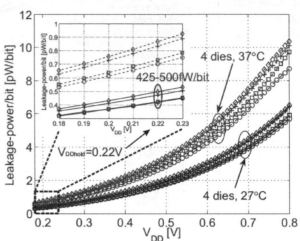

The measurements also show a tremendous impact of temperature on leakage power and the leakage variations across different dies due to process variations.

42   Comparison with Prior-Art Sub-V_T Memories

Benefits of designing 1 custom standard cell

- **Leakage power reduced by 50% (at no area increase) w.r.t. commercial standard cell latch [Meinerzhagen et al., JETCAS'11]**

Considered work: Full macro, measured, 65nm node

	[1]	[2]	[3]	[4]	This work
V_{DDmin} [mV]	380	250	700	350	**420**
V_{DDhold} [mV]	230	250	500	250	**220**
$E_{tot/bit}$ [fJ/bit]	54 (0.4V)	86 (0.4V)	-	55	**14 (0.5V)**
$P_{leak/bit}$ [pW/bit]	7.6 (0.3V)	6.1	6.0, 1.0[a]	-	**0.5**

[a] Leakage-power of bitcell only

[1] MIT: Calhoun and Chandrakasan, JSSC 2007; [2] MIT: Sinangil, Verma, and Chandrakasan, JSSC 2009; [3] Intel CRL: Wang et al., JSSC 2008; [4] STM: Clerc et al., ESSCIRC 2012

- **Lowest leakage-power/bit ever reported in 65nm CMOS**
- **Lowest active energy/bit-access ever reported in 65nm CMOS**
- **Reduce leakage in array and periphery!**

Overall, the conservative latch-based design allows the memory to operate and retain data reliably down to supply voltages that is far below the minimum supply voltage of other low-power memories reported in the literature.

43 Conclusions

- Low-power design is generally achieved through voltage scaling, at the expense of a reduced circuit operating frequency

- One of the main challenges, especially when operating at low voltages, is the impact of process variations which can require excessive design margins

- Adaptive body biasing allows to effectively reduce design margins by trading leakage for speed and vice versa even after production and at run-time

- Better correlation between sensor readings and circuit speed and close-loop adjustment of the bias values achieves the tightest margins, at the expense of area and power overhead

- Overall system power consumption is often dominated by memories, either directly through their own active or leakage power, or indirectly as memories limit the potential for voltage-scaling due to reliability concerns

- Replacing small memory macros with memories built from standard cells is a straightforward and powerful solution to reduce power consumption

CHAPTER 06

Low power Hall effect sensors. From design optimization to CMOS integration

Maria-Alexandra Paun

Ecole Polytechnique Fédérale
de Lausanne (EPFL), Switzerland

My name is Dr. Maria-Alexandra Paun and I am currently a Scientist at the Swiss Federal Institute of Technology in Lausanne, Switzerland. I am also the Chair of IEEE Switzerland Section and Chair of IEEE Switzerland Women in Engineering. It will be my pleasure to be delivering today my presentation entitled "Low power Hall effect sensors, in the framework of the doctoral course MICRO-622. From design optimization to CMOS integration". This is original work that has been developed during more than 7 years in the field of Hall effect sensors, including work during my PhD at EPFL, Switzerland (2009-2013) and postdoctoral research fellowships at University of Cambridge, UK (2013-2015) and published in various articles.

1

Main outline of the presentation

I shall start off my presentation by giving you the outline and the main points that will be discussed today. This course will guide you through an extensive study of the Hall effect sensors, which are a particular type of magnetic sensors. The course will provide an overview of the design

➤ This course will guide you through the study of Hall effect sensors, from design optimization to CMOS integration.

➤ It will provide design guidelines in CMOS.

➤ Both regular bulk and SOI Hall effect sensors have been integrated.

➤ Measurement results for their main parameters (sensitivity, offset) are provided.

➤ 3D physical and circuit models are developed for these sensors.

➤ Study of Hall mobility is also included.

➤ Overview on power consumption for the Hall cells is offered.

N.B. All the results presented here, obtained during more than 7 years work in Hall effect sensors, are original, published in reputed journals and belong to Maria-Alexandra Paun.

guidelines in CMOS technology, with illustration of various such integrated sensors in both regular bulk and SOI fabrication technological processes. To complete the presentation, measurement results are provided for the main parameters that govern the performance of these sensors, such as sensitivity and offset. Both three dimensional physical models and circuit models have been developed for the analysis of these sensors. A study of the Hall mobility is also included, both through theoretical and simulations perspectives. Additionally, an overview of the power consumption for the Hall cells under discussion is also included. All results presented here are original and belong to the author.

2

Uses of Hall effect sensors

I will be now focusing on presenting the uses of Hall effect sensors in the industry, with a focus on the automotive industry. As the statistics are mentioning, the automotive industry has the highest CAGR for the Hall sensors market projected for the next seven years (2017-2023). This means that this ever expanding market will

➤ Automotive industry has the highest CAGR for Hall sensors market (2017-2023).

➤ Current measurement is required for applications such as current control, protection of devices from overcurrent, and power management including control of motor drives, converter control, overcurrent protection, and battery management.

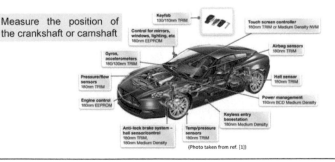

(Photo taken from ref. [1])

fuel the development of Hall effect sensors, which are to this day, a robust and cheap option for current sensing mass production. As we know, the current measurement is useful in the protection of the devices from overcurrent and in the power management.

Motor drives and battery management are important area in which these sensors are needed. In cars, a Hall sensors helps in measuring the position of the crankshaft and camshaft, as well as measure the wheel speed and sense the current in the wires.

3 Advantages of Hall effect sensors

Moving forward into the presentation, I shall now give an overview of the advantages of Hall effect sensors. They are reliable, reproducible and high performance sensors. Firstly, an output voltage signal is produced independently on the rate of the detected field. Secondly, this type of magnetic sensors are invariant to the ambient conditions, such as dust, humidity and vibrations. Thirdly, as these sensors do not have contact with any mechanical parts, it makes them sensitive but strong for the detection of movement. Moreover, the Hall effect sensors do not deteriorate in time and retain their quality. Also, they can measure large currents, while working in high temperature ranges. Last, but not least, they are attractive to the industry due to their cheap fabrication cots.

> An output voltage signal is produced independent of the rate of the detected field.
> They are invariant to ambient conditions, such as dust, humidity, and vibrations.
> They do not have contact with mechanical parts in their vicinity, rendering these sensors strong and sensitive enough to detect movement.
> They do not wear in time and retain quality and unlimited use.
> Their sensitivity depends on carrier mobility, which eliminates any perturbations due to surface elements and renders them reproducible and highly reliable.
> They can measure a large current and work in high temperatures ranges.
> Not lastly, they are cheap to fabricate.

4 Basic principles of Hall effect sensors

This slide is now focused on presenting the basic working principles of Hall Effect sensors. This type of sensors rely on the Hall effect principle, discovered by Edwin Hall at the end of the 19th century, more precisely in 1879. This states that if a current is flow through a conducting piece and the probe is placed under the influence of the magnetic field, the electrical carriers will be deviated under the action of Lorentz force and will create a non-zero tension drop known as Hall voltage. A classical Hall sensor is a symmetrical structure, it has a Greek-cross shape and is equipped with 4 contacts. There are two important equations that govern its performance, for the Hall voltage and the sensitivity. The absolute sensitivity depends on both physical and geometrical parameters. A very important aspect observed by myself during the PhD thesis that for devices in the same technological process a maximization of the geometrical correction G will allow for the highest performance.

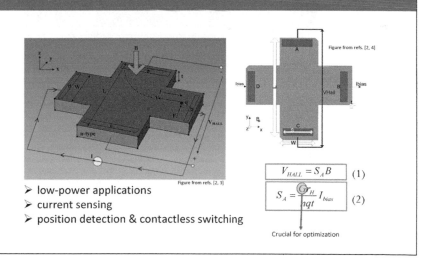

> low-power applications
> current sensing
> position detection & contactless switching

$$V_{HALL} = S_A B \quad (1)$$

$$S_A = \frac{G r_H}{n q t} I_{bias} \quad (2)$$

Crucial for optimization

5 Geometrical correction factor (I)

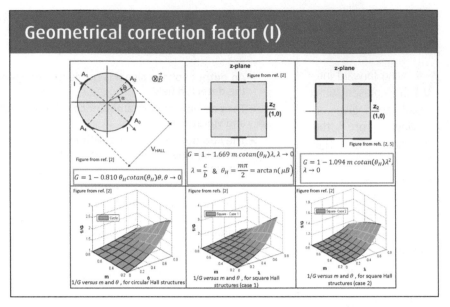

Now, more attention will be given to the analysis and evaluation of the geometrical correction factor (G) for various Hall cells. Circular Hall cells and two types of square (with contacts in the corners and with contacts on each side) Hall cells have been considered. To this purpose, the mathematical formulae for G are provided. They have been obtained through conformal mapping and give G as a function of the Hall angle (θ_H), angle θ, λ which is the ratio of the length of contacts to the length of the boundaries and m. Three-dimensional representations of 1/G with respect to m and θ have been included for the three Hall cells under discussion. In this way, the identification of the maximum geometrical correction factor is possible.

6 Geometrical correction factor (II)

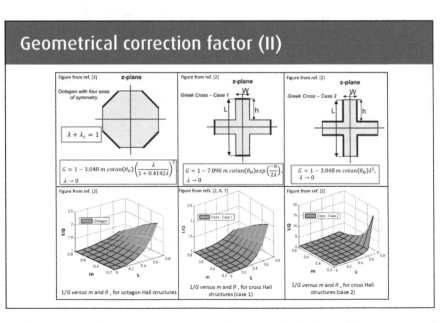

Further on, I shall talk about the geometrical correction factor (G) evaluation for another three Hall structures, namely the octagon with four axes of symmetry, Greek cross (case 1-contacts in blue are located on the width of the arms) and Greek cross (case 2-with complementary contacts). Again, the three dimensional representation is provided for 1/G, as a function of m and θ. It is important for the maximization of the Hall cells sensitivity, to have a good evaluation and optimization of the geometrical correction factor.

7

In this slide, a guiding procedure for the selection of length (L) and width (W) for rectangular Hall effect structures with small sensing contacts is provided. We start with such structures owing small contacts and then we look into the equation of the geometrical correction factor (G). We

Optimized dimensions selection

proceed to maximize this function G, by taking into account the proposed design specification of area (A) and sensing contacts length, s (dictated by the technological process selected). Ultimately, we obtain the dimensions of L and W that would guarantee the maximum G. We can then communicate these dimensions to the designer that would then provide a layout of the Hall cell with an informed decision of the dimensions for maximum sensitivity. Plots for L/W, W and Gmax vs. area and Gmax vs. s are provided. L/W and Gmax increase with the area, while Gmax decreases with the contacts length, s. We observe that for an imposed area A, L/W and G are maximum for minimum contacts length.

8

I will be now talking about the optimization of Hall cells design. Different cells (more than a dozen) were integrated in a regular bulk CMOs technological process. They are all symmetrical and orthogonal structures. As it has been found out, the geometry playing an important

Hall cells design optimization

➤ Different Hall cells were integrated in regular bulk CMOS technology.
➤ They are all symmetrical and orthogonal structures.
➤ The geometry plays an important role in the sensors performance.
➤ A powerful optimization tool is the analysis and optimization of *G*, according to Eq.(2).
➤ This allows for the sensitivity maximization of Hall cells under the same fabrication process.
➤ The geometrical correction factor for Greek-cross structure is given below w.r.t. L/W and Hall angle:

$$G \cong 1 - \frac{16}{\pi^2} \exp\left(-\frac{\pi}{2}\frac{L}{W}\right)\left[1 - \frac{8}{9}\exp\left(-\frac{\pi}{2}\frac{L}{W}\right)\right]\left(1 - \frac{\theta_H^2}{3}\right) \quad (3)$$

valid if $\quad 0.85 \le L/W < \infty \quad$ **and** $\quad 0 \le \theta_H \le 0.45$

role in the Hall effect sensors performance. As already pointed out, a powerful optimization tool is the analysis and optimization of the geometrical correction factor, according to Eq. (2). This procedure allows for the maximization of sensitivity of Hall cells under the same technological process. In Eq. (3), the geometrical correction factor (G) is given for Greek-cross structures with respect to the L/W ratio and Hall angle, valid for L/W greater than 0.85 and Hall angle between 0 and 0.45.

9

Integrated Hall cells main parameters

> ➤ Measurements results on nine different Hall effect sensors in regular bulk CMOS were obtained.
> ➤ Input resistance, sensitivity and offset drift can be found in the table below.

Objectives: offset @ T=300 K < ±30 µT & offset drift < ±0.3 µT/°C

Geometry Type	Basic	Low-doped	L	XL	45 Deg	Narrow Contacts	Borderless	Square	Optimum
Integrated Shape (CMOS 0.35 µm)									
R_p(kΩ) @ T=300 K, B=0 T	2.3	5.6	2.2	2.2	2.1	2.5	1.3	4.9	1.5
S_1 (V/T) @ I_{bias}=1 mA	0.0807	0.3392	0.0804	0.0806	0.0807	0.0822	0.0325	0.0884	0.0635
Offset drift (µT/°C) (4-phase current spinning)	0.409	0.067	0.264	0.039	0.373	0.344	0.526	0.082	0.328
L, W (µm) of the Active Area (N-well)	L=21.6 W=11.8	L=21.6 W=11.8	L=32.4 W=17.8	L=43.2 W=22.6	L=21.64 W=11.8	L=21.6 W=9.5	L=50 W=50	L=20 W=20	L=54 W=54
L/W	1.83	1.83	1.82	1.91	1.83	2.27	1	1	1
s (µm) for Sensing Contacts	11	11	16	20.7	11	1.5	2.3	2.3	5.4
Geometrical Correction Factor (G)	0.913	0.913	0.912	0.924	0.913	0.87	0.76	0.73	0.74

At this point, my talk is focused on giving you a characterization of nine different Hall cells that have been integrated in regular bulk CMOS process and thoroughly measured. Their main parameters, such as input resistance, sensitivity and offset drift have been included. All these shapes have been designed with very strict specification of the project, namely an offset at room temperature less than ±30 µT and an offset temperature drift less than ±0.3 µT/°C. Structures like basic, L, XL, 45 degrees, narrow contacts, square, borderless and optimum have been integrated. The L and XL Hall cells are scaled versions (with 1.5 and 2 factor respectively) of the basic cell. As expected, the sensitivity of the scaled cells is the same. Moreover, we can easily notice that the XL Hall cell provides the lowest offset drift, 0.039 µT/°C, which is a few times better that the imposed specifications. This is explained by the larger the cell, the less effect of the errors on borders in the offset.

10

AC and Dc measurements

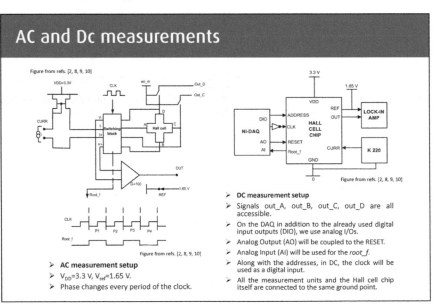

Figure from refs. [2, 8, 9, 10]

Figure from refs. [2, 8, 9, 10]

> ➤ AC measurement setup
> ➤ V_{DD}=3.3 V, V_{ref}=1.65 V.
> ➤ Phase changes every period of the clock.

> ➤ DC measurement setup
> ➤ Signals out_A, out_B, out_C, out_D are all accessible.
> ➤ On the DAQ in addition to the already used digital input outputs (DIO), we use analog I/Os.
> ➤ Analog Output (AO) will be coupled to the RESET.
> ➤ Analog Input (AI) will be used for the *root_f*.
> ➤ Along with the addresses, in DC, the clock will be used as a digital input.
> ➤ All the measurement units and the Hall cell chip itself are connected to the same ground point.

To fully characterize the behaviour and performance of the integrated Hall cells, extensive measurements have been carried out. To this purpose, both AC and DC measurement setups have been developed. The proposed AC measurement setup on the Hall cell provides as output signal either the Hall voltage (if the probe is place in magnetic field of strength B) or the averaged residual offset (if no magnetic field). VDD=3.3 V and VREF=1.65 V. The phase changes every period of the clock. The CLK frequency is 37 Hz, while the demodulation is half of this. For the DC setup, individual information on each phase can be obtained. On the DAQm we use both digital inputs/outputs (DIO) and analog inputs-outputs (AO, AI). Along with the addresses, in DC the clock will be used as digital input. We have used Keithley K 220 for current generation and a lock-in amplifier.

11

Techniques to cancel the offset

O ne of the major drawbacks of a Hall cell is its offset, which is non-zero. This is an undesired, parasitic effect that adds to the Hall voltage to create the output voltage. By offset we understand the voltage signal, in the absence of magnetic field. This is due to imperfections in the fabrication process,

> Cell polarization for different shapes

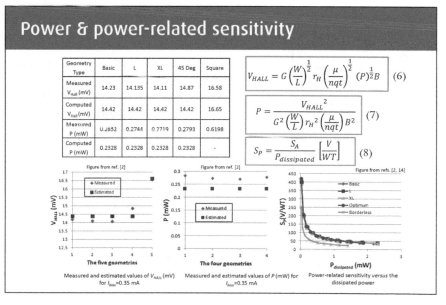

Greek-cross cell polarization
Figure from refs. [2, 3, 11, 12]

Borderless cell polarization
Figure from refs. [2, 3]

Optimum cell polarization
Figure from refs. [2, 3, 13]

> Single phase offset and residual offset

$$V_{out} = V_{HALL}(B) + V_{offset} \quad (4)$$

$$Offset_{residual\ (4\ phase)} = \frac{V_{P1} - V_{P2} + V_{P3} - V_{P4}}{4} \quad (5)$$

Phases	I_{bias}	V_{HALL}
Phase 1	a to c	b to d
Phase 2	d to b	a to c
Phase 3	c to a	d to b
Phase 4	b to d	c to a

misalignments of the contacts and etching masks, etc. The most widely used technique to cancel the offset is the current-spinning technique or connection-commutation. This concept is based on the fact that by rotating the Hall cell, the offset voltage changes sign, while the Hall voltage does not. This means that the Hall cells will be rotated a number of times (2 or 4-phases are common) and the resultant offset will be averaged and therefore decreased. The Hall cell polarizations of different structures (Greek-cross and square type) are also presented.

12

Power & power-related sensitivity

T his part of the talk is now focused on presenting the power, power-related sensitivity and Hall voltage for five different Hall cells (Basic, L, XL, 45 Deg and Square). The formula for the Hall voltage as function of the power is also provided. From this formula, the dissipated power is extracted. Plots showing the Hall voltage

Geometry Type	Basic	L	XL	45 Deg	Square
Measured V_{Hall} (mV)	14.23	14.135	14.11	14.87	16.58
Computed V_{Hall} (mV)	14.42	14.42	14.42	14.42	16.65
Measured P (mW)	0.2852	0.2744	0.2719	0.2793	0.6198
Computed P (mW)	0.2328	0.2328	0.2328	0.2328	-

$$V_{HALL} = G \left(\frac{W}{L}\right)^{\frac{1}{2}} r_H \left(\frac{\mu}{nqt}\right)^{\frac{1}{2}} (P)^{\frac{1}{2}} B \quad (6)$$

$$P = \frac{V_{HALL}^2}{G^2 \left(\frac{W}{L}\right) r_H^2 \left(\frac{\mu}{nqt}\right) B^2} \quad (7)$$

$$S_P = \frac{S_A}{P_{dissipated}} \left[\frac{V}{WT}\right] \quad (8)$$

Figure from ref. [2]

Measured and estimated values of V_{HALL} (mV) for I_{bias}=0.35 mA

Figure from ref. [2]

Measured and estimated values of P (mW) for I_{bias}=0.35 mA

Figure from refs. [2, 14]

Power-related sensitivity versus the dissipated power

and dissipated power for the considered geometries are also included. A new metric called power-related sensitivity, SP, defined as the ratio between the absolute sensitivity and the dissipated power is introduced. The plot of the power-related sensitivity versus the dissipated power is also provided. We can notice that the power-related sensitivity is the smallest for the borderless cell, while it is higher and almost equal for the other four shapes.

13   Offset measurement and analysis (I)

> ➢ Single phase offset was measured by an automated measurements setup
> ➢ Quadratic behaviour of the single phase offset with the biasing current was proven

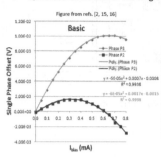

Figure from refs. [2, 15, 16]

Coefficients	Basic Cell	
	Phase P2	Phase P3
a (V/mA)	0.0007	0.00017
b(V/m²A²)	-5E-5	-6E-5
R²	0.9938	0.9998

Measured single phase offset (V) versus I_{bias} for Basic Hall cell

I shall now talk about the offset measurement and analysis. As already pointed out, the offset is a parasitic effect and we would like to have it as small as possible. To the purpose of offset evaluation for the Hall cells, the single phase offset was measured by an automated measurement setup. The quadratic behaviour of the single phase offset with the biasing current was proven and the corresponding coefficients were extracted. This information is necessary if compensation blocks for the temperature behaviour are added. The measured single phase offset variation for a biasing current in the interval 0 to 0.8 mA is also presented, for the Basic Hall cell.

14   Offset measurement and analysis (II)

> ➢ Magnetic equivalent offset of the Hall cells was measured using an automated setup on 64 cells at a time

$$B_{offset} = \frac{V_{offset}}{S_A} \quad (9)$$

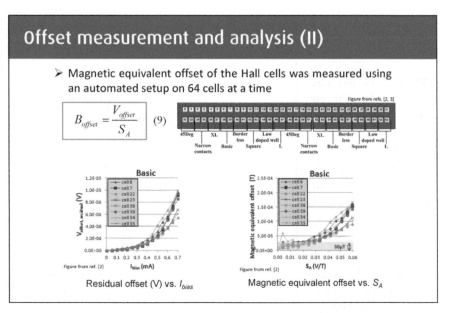

Figure from refs. [2, 3]

Residual offset (V) vs. I_{bias}

Magnetic equivalent offset vs. S_A

To continue the evaluation of the Hall cells offset, the magnetic equivalent offset of the Hall cells was measured using an automated measurement setup able to test 64 cells at a time. The magnetic equivalent offset is defined as the ratio of the offset voltage to the absolute sensitivity and is measured in Tesla (T), as it can be seen in Eq. (9). On the chip with 64 cells, each cell is positioned 8 times, in the sequence depicted above. The advantages of using this setup is that we obtain accurate and fast information on the offset value. The measured residual offset voltage versus the current and the magnetic equivalent offset versus the absolute sensitivity are also included, for the Basic Hall cell, tested 8 times on the chip. At this point, we mention that the magnetic equivalent residual offset is not a direct function of the sensitivity, but an implicit one via the biasing current.

15 Offset measurement and analysis (III)

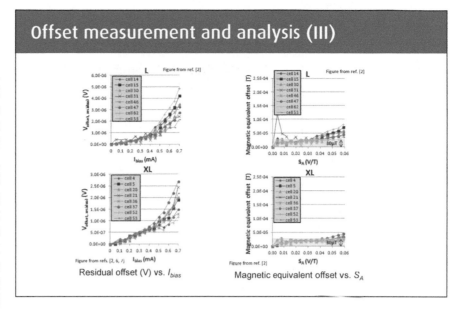

Residual offset (V) vs. I_{bias} Magnetic equivalent offset vs. S_A

Further on, measurement results regarding the offset of Hall cells is presented for L and XL cells respectively, both for residual offset voltage versus biasing current and magnetic equivalent offset versus the absolute sensitivity. The sensitivity of the Hall cells was measured under a magnetic field strength given by a permanent magnet of B=0.5 T. We can immediately notice that the XL provides the minimum offset which fits the best in the band of specification of 30 µT at room temperature. This data has also been obtained using the automated measurement setup presented before, and each Hall cells was tested 8 times. Again the residual offset is an average of 4 phases.

16 Offset measurement and analysis (IV)

➤ **Quadratic behaviour of the 2-phase residual offset with biasing current was investigated**

$$V_{residual,2-phase\,offset} = V_{offset,E} + K_{residual\,offset} I^2 \qquad (10)$$

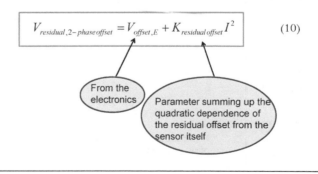

From the electronics

Parameter summing up the quadratic dependence of the residual offset from the sensor itself

At this point, we continue with the offset analysis. We will talk about the quadratic behaviour of the two-phase residual offset with the biasing current. The linear part of the formula in Eq. (10) is given by the electronics, while the second order behaviour has a residual offset parameter summing up the quadratic dependence of the residual offset from the sensor itself. Even though both the two-phase and four-phase residual offset (as it could be seen on the next slide). It is worth mentioning that they have different contributing terms. While the two-phase residual offset voltage has a free term and second order term, as we will see next, the four-phase residual offset voltage is composed of a first order and second order term. These assumptions have been validated by measurement results, presented all throughout my talk.

17

Offset measurement and analysis (V)

➢ **Quadratic behaviour of the 4-phase residual offset with biasing current**

$$V_{residual, 4-phase offset} = aI + bI^2 \qquad (11)$$

Figure from refs. [2, 15]

the bias lead is influenced by the magnetic field

contributed by the thermoelectric signal caused by a temperature difference* between the contacts of the Hall plate.

*In the bias lead, this temperature difference is explained by Joule heating.

Measured 4-phase residual offset (V) versus I_{bias} for basic Hall cell, fitted with second order function

Further on, the quadratic behaviour of the four-phase residual offset with the biasing current was emphasized. In Eq. (11), giving the formula for the residual offset voltage we can identify two terms. The first order contribution comes from the fact that the bias lead is influenced by the magnetic field. Secondly, the second order term is contributed by the thermoelectric signal caused by a temperature difference between the contacts of the Hall plate. Moreover, in the bias lead, this temperature difference is explained by the Joule heating effect.

The four-phase residual offset voltage measured against the biasing current, for the Basic Hall cell, proves this second order behaviour. The biasing current was chosen from 0 to 1 mA.

18

Offset measurement and analysis (VI)

➢ **Encapsulation offset was measured, for the Hall cells**

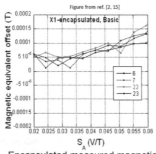

Figure from ref. [2, 15]

➢Offset of the non-packaged cells could be affected by the packaging (through strain, pressure, etc.)

➢It is recommended to also look at the post-encapsulation offset of the Hall sensors.

➢Tested several times on the same chip, SOP 16 packaging was used

Encapsulated measured magnetic equivalent residual offset (T) versus absolute sensitivity for Basic Hall cell

We were not only interested about the offset of the naked Hall cells, but most importantly about the offset of the encapsulated Hall cells. As we know, the offset levels are sensitive and could be affected by the packaging. It can increase due to piezo-electric effect, pressure or mechanical strain. Therefore, it is crucial to see if the offset levels are kept in the specifications, for the encapsulated devices. We must also look at the post-encapsulation offset of the Hall cells. To this purpose, we have tested several times on the same chips and for many chip, encapsulated samples of Hall cells. The SOP 16 packaging was chosen for encapsulation. We also present a plot of the measured magnetic equivalent offset plotted against the absolute sensitivity for the Basic Hall cell, in chip X1.

19

Offset measurement and analysis (VII)

Regarding the offset, not only do we need to investigate its behaviour and values at room temperature, but also the temperature drift of the offset is important to measure and quantify. As the Hall effect sensors are used in various low-power applications in the industry and in the cars, an important

➢ Magnetic equivalent offset temperature variation was investigated
➢ Measured for Basic Hall cell in regular bulk CMOS

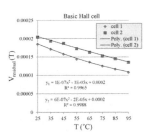

Measured magnetic equivalent offset (T) versus the temperature, for the Basic Hall cell

➢ Offset temperature drift is important.

➢ Temperatures from 25 to 95 °C were considered, by the means of a Temptronic oven.

➢ The same Basic cell was tested twice.

Averaged Offset Drift for Basic Hall cell: **0.409 μT/ °C**

temperature cycle will always be operating. The magnetic equivalent offset temperature variation was therefore investigated. Measurements for the Basic Hall cell in regular bulk CMOS are offered. A temperature range from 25 °C to 75 °C was considered and provided by a Temptronic oven in the laboratory measurements. The same Basic Hall cell was tested twice. The calculated averaged offset drift for the Basic Hall cell was in the order of 0.40 μT/°C.

20

Hall mobility study

This part of the talk is now devoted to the study of Hall mobility in Hall cells fabricated in two different technological processes (CMOS regular bulk and SOI). As we know, for a batter control of the performance of Hall cells, an evaluation of the Hall mobility is required. This work evaluates the Hall mobility with a focus

➢ This work evaluates the Hall mobility of Hall cells in two different technological processes (bulk and SOI).

➢ The focus of this work is on the Basic Hall cell in the two processes and 3D models are offered.

➢ A theoretical computation as well as simulation results are included for Hall mobility evaluation.

➢ The most important parameters of the Hall cells are computed through the 3D model and also measured experimentally.

on the Basic Hall cell in the two above mentioned process, by doing both a theoretical analysis and through results obtained by three-dimensional physical model. The most important parameters of the Hall cells under discussion are computed through the developed three-dimensional model and also measured experimentally.

21 — Three-dimensional models (I)

➤ The Basic Hall structure follows the XFAB XH 0.35 fabrication process.

➤ *active n-well region:* **Arsenic doping**
  ✓ **Gauss profile implantation**
  ✓ **$1.5*10^{+17}$ cm^{-3}**

Parameter	Value
p-substrate (µm²)	27*27
L (µm)	21.6
W (µm)	9.5
Contacts s (µm)	8.8

Cell BASIC

Figure from refs. [2, 14, 15]

➤ *electrical contacts*

The 3D model of regular bulk Basic Hall cell

➤ *p-substrate:* **Boron doping concentration of 10^{+15} cm^{-3}**

An important tool in the performance assessment of Hall cells is the development of three-dimensional physical models. The simulations can guide designers when integrating new Hall cells and their results could allow for the optimum shape selection in terms of the necessary performance. The TCAD Sentaurus of Synopsys simulator is an excellent instrument to achieve this objective. It solves the carrier transport in semiconductor devices under magnetic field influence, by computing the Poisson equation and holes and electrons continuity equations. The actual technology used for real sensors integration was reproduced in the Hall cells construction by computer simulation. You can find all the model parameters and their numerical values in this slide. By carefully adapting the meshing strategy for a good tradeoff between results accuracy and simulation run time, different Hall cells were modelled and simulated.

22 — Three-dimensional models (II)

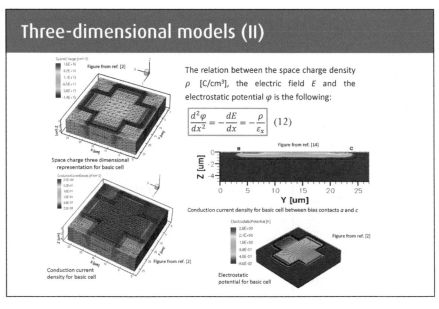

The relation between the space charge density ρ [C/cm³], the electric field E and the electrostatic potential φ is the following:

$$\frac{d^2\varphi}{dx^2} = -\frac{dE}{dx} = -\frac{\rho}{\varepsilon_x} \quad (12)$$

Space charge three dimensional representation for basic cell

Conduction current density for basic cell between bias contacts *a* and *c*

Conduction current density for basic cell

Electrostatic potential for basic cell

The exact XFAB CMOS fabrication process parameters are not provided by the foundry. We have obtained the necessary numerical values for our model through parameter extraction from measurements and numerical computations, such that to allow for a precise reproduction through the three-dimensional model of the integrated cells. For the regular bulk CMOS process, we have used a p-substrate with a Boron concentration of 10+15 cm-3 and an active n-well region doped with Arsenic (Gauss profile with peak concentration of 1.5·10+17 cm-3). This doping profile allows for an average mobility of 0.0630 m2V-1s-1 at the surface of the Hall cells. The thickness is 5 µm for the p substrate and 1 µm for the implantation of the n-doped profile active region respectively. For the simulated Basic Hall cell, three-dimensional representation of the space charge, conduction current density and electrostatic potential have been obtained.

23

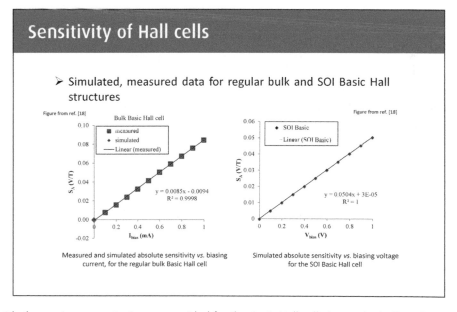

Three-dimensional models (III)

➢ The Basic Hall structure now follows the XFAB XI10 SOI fabrication process.

Figure from refs. [11-13, 18-19]

Figure from ref. [18]

SOI XFAB XI10 fabrication layers

DOPING CONCENTRATIONS OF THE SOI PROCESS

Layer	Type	Numerical value (cm⁻³)	Depth (mm)
Wafer (handle) substrate	Si, p-doped (Boron)	6.5E+14	3
Dielectric	Buried SiO₂	N/A	1
p-substrate in active Silicon layer	Si, p-doped (Boron)	1E+15	0.25
n-well in active Silicon layer	Si, n-doped (Arsenic)	5E+16	0.125

Now, we shall be talking about the three-dimensional physical model of the CMOS Silicon on Insulator (SOI) Hall cells counterparts. The SOI technology advantages are increased sensitivity (due to shallower implantation profile of the active area), increased radiation resistance, lower leakage current through the dielectric. More precisely, the Basic Hall cell now follows the non-fully depleted XFAB XI10 SOI fabrication process. In this slide, you have an overview of their geometrical and fabrication parameters (values of the doping concentrations in the implantation profiles). The depth of the active Silicon layer is 0.25 µm, while the buried oxide underneath has a thickness of 1 µm. Meshed structures contain a sufficient amount of points for a good trade-off between simulation run time and accuracy.

24

Sensitivity of Hall cells

➢ Simulated, measured data for regular bulk and SOI Basic Hall structures

Figure from ref. [18]

Figure from ref. [18]

Measured and simulated absolute sensitivity vs. biasing current, for the regular bulk Basic Hall cell

Simulated absolute sensitivity vs. biasing voltage for the SOI Basic Hall cell

An important performance indicator of Hall cells is their sensitivity. We have both absolute sensitivity and relative sensitivity (current-related or voltage-related). The absolute sensitivity of Hall cells is given by the ratio of the measured Hall voltage to the magnetic field strength (B). It is inversely proportional to the thickness (t) of the implantation profile and with the carrier concentration (n) in the active region. Also, as already mentioned, for Hall cells under the same fabrication process, it is the maximization of the geometrical correction factor (G) which is necessary for achieving the highest sensitivity. In this slide, both simulated and measured data are provided for the Basic Hall cells in regular bulk and SOI technological processes. For the bulk Basic Hall cell, the absolute sensitivity has been plotted against a biasing current from 0 to 1 mA, while for the SOI structure, its sensitivity is given for various biasing voltages in the interval from 0 to 1 V.

25 · Circuit model of Hall cells (I)

Now we move into my talk's part dedicated to present the circuit model of the Hall cells. This circuit model was developed by the author in Cadence and entirely coded in Verilog-A. It is a Finite Element Model (FEM), based on an elementary cell (e). The elementary cell is then reproduced several times to recreate the needed

- Circuit model developed in Cadence and coded in Verilog-A
- FEM lumped circuit model based on an elementary cell

$$R_X = \frac{L}{tW\sigma}, \; R_Y = \frac{W}{tL\sigma} \quad (13)$$

$$i_X = \frac{W}{L}\mu_H B i_Y, \quad i_Y = \frac{L}{W}\mu_H B i_X \quad (14)$$

$$i_X = K_{XY} i_Y, \quad i_Y = K_{YX} i_X \quad (15)$$

$$K_{XY} = \frac{W}{L}\mu_H B, \quad K_{YX} = \frac{L}{W}\mu_H B \quad (16)$$

Figure from ref. [20]

shape of the Hall cell. The elementary cell is based on parametrized resistances (RX, RY), current-controlled current sources (F1, F2). The magnetic and electric paths are separated. The current flowing through the branch x is controlled by the current flowing through the opposite branch y, through a factor K. Zero voltage DC sources are additionally placed on the branches in order to sense the current flowing through them. The elementary cell is the unit of the FEM model and it contains 8 ports.

26 · Circuit model of Hall cells (II)

As mentioned before, the elementary cell is repeated several times in order to recreate the Hall cell shape to be analyzed. A sufficient number of elementary cell should be chosen in order to guarantee for a good management of the interconnections, boundary conditions and current flow path. In this slide, one can

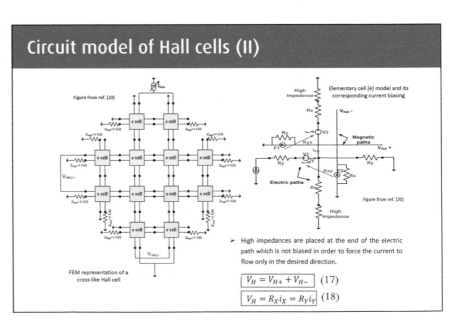

$$V_H = V_{H+} + V_{H-} \quad (17)$$

$$V_H = R_X i_X = R_Y i_Y \quad (18)$$

see the FEM model based on the elementary cell for the cross-like Hall cell. In a similar way, other shapes such as square could be represented. A current polarization with Ibias is shown in this slide, with high impedances Zhigh=1 GΩ placed at the end of the electric field path which is not biased, in order to force the current to flow only in the desired direction. The calculation of the Hall voltage comes from the difference in the Hall voltage on the "plus" branch and the Hall voltage on the "minus" branch. In this slide, a polarization of a single elementary cell is also provided.

27

Circuit model of Hall cells (III)

Polarization of a non-homogeneous Greek-cross cell for offset analysis

Figure from ref. [22]

Figure from ref. [2]

Misalignment offset analysis

Simulated offset voltage (V) *versus* I_{bias} for four asymmetries

➤ The parameterized resistances used to model the different cells responsible for inducing asymmetries are rewritten:

$$R_Y = \frac{W\cos\alpha}{tl\sigma} \quad (19)$$

Asymmetry	Expression
a_1	$\cos5° = 0.9993$
a_2	$\beta=0.8$
a_3	$\beta=0.7$
a_4	$\beta=0.6$

Further on, we shall be speaking about the use of the freshly presented FEM lumped circuit model for Hall cells offset evaluation. In order to emulate the offset, various different cells, depicted here in red and named "a", will be added in order to break the symmetry. One can see in this slide, the polarization of a non-homogeneous Greek-cross cell for the offset analysis. Inside the new cell "a", included for offset emulation, a different parametrized resistance can be used, according to Eq. (19). In this slide, four different asymmetries are presented in the Greek-cross Hall structure and the misalignment offset analysis provided numerical evaluation of the voltage offset in these four cases. To obtain the four asymmetries, in Eq. (19), different values for the cosα are used, which will create different parameterized resistances and therefore imbalance in the Hall cell, for offset evaluation. The obtained values are in good agreement with the measured data for single phase offset.

28

Hall mobility evaluation (I)

➤ Parameter investigated both through analytical computation and 3D simulations

$$\mu_H = r_H\mu \quad (20)$$

$$\mu_H = -\frac{V_{HALL}}{\rho B}\frac{t}{I} \quad (21)$$

The following is a transcendental eq.

$$\frac{V_{HALL}}{I}\frac{t}{\rho} = \mu_H - 1.045\arctan(\mu_H B)\exp(\frac{-\pi h}{W}), \frac{h}{W} \to \infty \quad (22)$$

➤ The Masetti formula is used in simulations:

$$\mu_{dop} = \mu_{min1}e^{-\frac{P_C}{N_{A,0}+N_{D,0}}} + \frac{\mu_{const} - \mu_{min2}}{1+(\frac{N_{A,0}+N_{D,0}}{C_r})^\alpha} - \frac{\mu_1}{1+(\frac{C_s}{N_{A,0}+N_{D,0}})^\beta} \quad (23)$$

where μ_1, μ_{min1} and μ_{min2} are reference mobilities, C_s, C_r and P_C are the reference doping concentrations. Additionally, α and β are two coefficients, specific to silicon, whose values are $\alpha=0.680$ (for electrons), $\alpha=0.719$ (for holes) and $\beta=2$ (for both holes and electrons).

Moving further into my presentation, I shall devote now time to talk about the Hall mobility evaluation in Hall cell. This very important parameter has been investigated both through analytical computation and 3D physical simulations. In Eqs. (20) and (21), one could see the relation between Hall mobility and regular carrier mobility, as well as between Hall mobility and Hall voltage. In Eq. (22), a transcendental equation which gives the Hall mobility is also given. In the three-dimensional model of the Hall cell, a specific formula, called the Masetti formula in Eq. (23) is used for Hall mobility evaluation. This takes into account the following parameters: μ1, μmin1 and μmin2 are reference mobilities, Cs, Cr and PC are the reference doping concentrations. Additionally, α and β are two coefficients, specific to silicon, whose values are α=0.680 (for electrons), α=0.719 (for holes) and β=2 (for both holes and electrons).

29

After developing the three-dimensional model of the Hall cell and including the necessary equations in the Physics section of the simulations files, we can then proceed to a numerical evaluation of the Hall mobility. The magnetic field strength B=0.5 T and the mesh has been optimally refined to allow for precise assessment of the Hall mobility as different location points in the 3D structure. In this slide, we will be presenting data obtained for the Hall mobility obtained at the surface of the device, for the SOI Basic Hall cell. The first plot, to the left-hand side, shows the Hall mobility for SOI Basic Hall cell versus the Y dimension, for an Ox cut X1=13.37

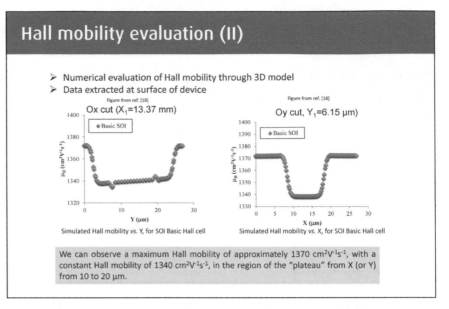

µm. Secondly, in the right-hand side, we can see simulation results for the Hall mobility versus the X dimension, obtained for an Oy cut Y1=6.15 µm. We observe a maximum Hall mobility of approximately 1370 cm2V-1s-1, with a constant Hall mobility of 1340 cm2V-1s-1, in the region of the "plateau" from X (or Y) from 10 to 20 µm.

30

In this slide, we shall talk about the Hall mobility evaluation for the Hall cells in regular bulk technology process. The plot presents the Hall mobility plotted against X dimension, for Oy cut Y=13.15 µm, for regular bulk Hall cell. The same equations presented in the slides above have been used in the model for Hall

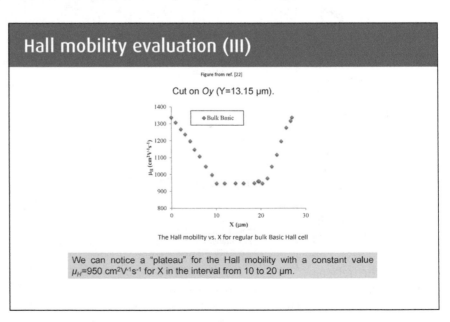

mobility analysis and numerical assessment. We can notice a "plateau" for the Hall mobility with a constant value µH=950 cm2V-1s-1 for X in the interval from 10 to 20 µm. We mention that these are simulation results obtain using TCAD Sentaurus from Synopsys. The same type of analysis can be extended to other Hall cells.

31

Conclusions (Part I)

➤ This part was intended to evaluate the Hall mobility for Hall cells in two different technological processes.

➤ An emphasis on the Basic Hall cell was preferred, with accurate 3D models and doping profiles being provided.

➤ A theoretical computation as well as numerical evaluation through 3D physical simulations was used for Hall mobility evaluation.

➤ Maximum and minimum values of the Hall mobility as well as the "plateau" region were provided in both technological processes.

The moment had come to now draw the conclusions of Part I of my talk. This part was intended to evaluate the Hall mobility for Hall cells in two different technological processes. An emphasis on the Basic Hall cell was preferred, with accurate three-dimensional models, with precise dimensions and doping profiles being provided. A theoretical computation as well as numerical evaluation through 3D physical simulations was used for Hall mobility evaluation, for the Hall cells in both regular bulk and SOI technological processes. Maximum and minimum numerical values of the Hall mobility as well as the "plateau" regions were provided in both technological processes.

32

Effect of Hall cells scaling

➤ Different Hall cells have been integrated in regular bulk CMOS technology (XFAB CMOS XH 0.35 μm) and analyzed in terms of their specific parameters.

➤ Geometry plays an important role in Hall cells performance. The focus of this work is on three Greek-cross Hall cells.

➤ Effect of scaling on the performance has been analyzed.

➤ Accurate single and multi-phase offset numerical estimations have been provided for the Basic Hall cell.

➤ The most important parameters of the Hall cells, based on regular bulk structure, are also evaluated through three-dimensional physical simulations.

In the slides at the beginning of my presentation, I was showing the many different Hall cells that we have fabricated and tested. These Hall cells have been integrated in regular bulk CMOS technology (XFAB CMOS XH 0.35 μm) and analyzed in terms of their specific parameters. For the Greek-cross cells, I have introduced the basic, L and XL cells. The L and XL cells are scaled up versions of the basic Hall cell, with a scale factor of 1.5 and 2 respectively. Geometry plays an important role in Hall cells performance. The focus of this work is on three Greek-cross Hall cells. Effect of scaling on the performance has been analyzed. Accurate single and multi-phase offset numerical estimations have been provided for the Basic Hall cell. The most important parameters of the Hall cells, based on regular bulk structure, are also evaluated through three-dimensional physical simulations.

33

Study of the Greek-cross Hall cells

➢ Three Greek-cross cells have been considered, with progressive scaled up dimensions

Type of cell	Geometrical parameters		
	Length, L (µm)	Width, W (µm)	Contacts length s, (µm)
Basic	21.6	9.5	8.8
L	32.4	14.25	13.55
XL	43.2	19	18.3

➢ Hall Cells Performance Summary

	Measured parameters		
	Structure	S$_A$(V/T)	R(kΩ)
Basic		0.08	2.3
L		0.08	2.2
XL		0.08	2.2

As already mention, three Greek-cross cells have been considered, with progressive scaled up dimension. We would like to see the benefit of the scaling on the Hall cells performance and offset decrease. In this slide, you can see in the first table a detailed presentation of the dimensions (length, L, width, W and contacts length, s) for the three Hall cells considered. It is obvious that L/w ratio stays the same. These three Hall cells have been thoroughly analyzed and electrically and magnetically tested (B=0.5 T). In the second table, one can see the performance summary, containing measured main parameters of the Hall cells under discussion. As the absolute sensitivity depends on L/W, it will obviously not change the three Hall cells, and we obtain S=0.08 V/T. an input resistance of 2.3 Ω was obtained for Basic Hall cell, while for L and XL Hall cells, R=2.2 Ω.

34

Scaling influence on Hall cells offset

➢ Residual offset obtained by AC setup
➢ First and second order coefficients of the residual offset variation with the biasing current are extracted
➢ Offset decreases with 66% for L and 80% for XL, w.r.t. basic cell

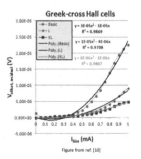

Figure from ref. [10]

$$V_{offset, residual} = aI + bI^2 \quad (24)$$

Values of coefficients	Basic	L	XL
a (V/mA)	-E-5	-4E-6	-E-6
b(V/m^{2}A^2)	3E-5	E-5	6E-6
R^2	0.9869	0.9708	0.9807

For the three Hall cells, basic L and XL, measurements were performed to reveal the scaling influence on the Hall cell residual offset value. As we already mentioned, the author found out that the larger you do the Hall cells, the less effect of the border errors on the offset. This assumption has been validated through extensive measurements. The residual offset was obtained using the automated measurement setup. In the provided plot, the measured four-phase residual offset versus the biasing current from 0 to 1 mA has been given. In Eq. (24), we remember the parabolic behaviour of the residual offset voltage with the biasing current. First and second order coefficients of the residual offset variation with the biasing current are extracted. We have observed that the offset decreases with 66% for L and 80% for XL, with respect to the basic cell. Therefore, we can say that a considerable decrease in the offset values is obtained for the XL cell, which is the basic cell scaled twice.

35

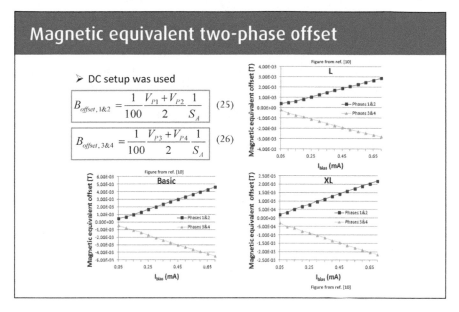

Magnetic equivalent two-phase offset

> DC setup was used

$$B_{offset,1\&2} = \frac{1}{100} \frac{V_{P1}+V_{P2}}{2} \frac{1}{S_A} \quad (25)$$

$$B_{offset,3\&4} = \frac{1}{100} \frac{V_{P3}+V_{P4}}{2} \frac{1}{S_A} \quad (26)$$

Now we shall see how the two-phase residual magnetic equivalent offset is influenced by the scaling of the Greek-cross Hall cells. This time, the developed DC measurement automated setup was used to retrieve the information on each single phase offset value of the Hall cell. Then, by using the formulae in Eqs. (25) and (26), the two-phase magnetic equivalent residual offset is calculated for both phases 1 and 2 and phases 3 and 4 respectively. Three plots are provided at this point, which show the measured two-phase residual magnetic equivalent offset in T, for the basic, L and XL Hall cells. We can see the numerical values and then the offset levels for each structure. Again, as in the case with the four-phase residual offset, there is a significant improvement in the offset levels for the XL Hall cell with respect to the other basic and L Hall cells, both in the cases of 1&2 phases and 3&4 phases.

36

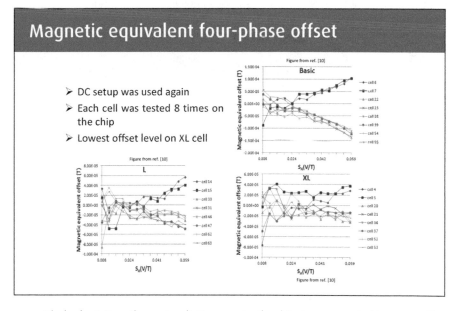

Magnetic equivalent four-phase offset

> DC setup was used again
> Each cell was tested 8 times on the chip
> Lowest offset level on XL cell

While in the previous slide, we were concerned with the effect of the Hall cells scaling on the magnetic equivalent two-phase residual offset, now we look into the same type of analysis but for the magnetic equivalent four-phase residual offset. Again the DC automated measurement setup was used. Three plots are provided, depicting the measured magnetic equivalent four-phase residual offset versus the absolute sensitivity for the three Greek-cross Hall cells with the progressive scaling (basic, L and XL). As we mentioned before the presented function is not a direct function, but an implicit one via the biasing current. However, the values of the offset are worth looking at for various intervals of the sensitivity. Each cell was tested eight times on the same chip. We notice that once again the lowest offset level belongs to the XL cell.

37 Conclusions (Part II)

My presentation has now reached the point of providing the conclusions of the second part of my talk. This work was intended to analyse the behaviour of Hall cells in regular bulk CMOS fabrication process. The effect of scaling on performance was also analyzed, by looking at the basic, L and XL Greek-cross Hall cells. Different offset numerical estimations (single phase, multi-phase offset voltage, residual offset, magnetic equivalent offset, post-encapsulation offset) were investigated. It was observed that XL Hall cell provided the minimum offset levels among all integrated cells (and a few times better than the project specifications in Slide 10) and it was chosen as the optimum candidate for the second integration with the ASIC electronics. Special attention was given to prove the quadratic behaviour of the offset with the biasing current. The variation of the magnetic equivalent offset with the temperature was also performed for the CMOS Basic Hall cell, for a 25 to 95 °C range.

> This work was intended to analyze the behaviour of Hall cells in regular bulk CMOS fabrication process.

> The effect of scaling on performance was also analyzed.

> Different offset numerical estimations (single phase, multi phase offset voltage, residual offset, magnetic equivalent offset, post-encapsulation offset) were investigated.

> Special attention was given to prove the quadratic behaviour of the offset with the biasing current.

> The variation of the magnetic equivalent offset with the temperature was also performed for the CMOS Basic Hall cell, for a 25 to 95 °C range.

38 References

1. Synopsys website: https://www.synopsys.com/

2. Paun, M.A., "Hall cells offset analysis and modeling approaches", PhD thesis, EPFL, Switzerland, 2013.

3. Paun, M.A., Sallese, J.M., Kayal, M., "Hall Effect Sensors Design, Integration and Behaviour analysis", *Journal of Sensors and Actuator Networks*, Vol. 2, Issue 1, 2013, pp. 85-97.

4. Paun, M.A., Sallese, J.M., Kayal, M., "Geometry influence on the Hall effect devices performance", *U.P.B. Scientific Bulletin, Series A: Applied Mathematics and Physics*, Vol. 72, Issue 4, 2010.

5. Paun, M.A., Paun, V.A., "On the geometrical optimization of CMOS Hall cells, rectangular and square", *U.P.B. Scientific Bulletin, Series A: Applied Mathematics and Physics*, Vol. 78, Issue 4, 2016, pp. 329-340.

6. Paun, M.A., Sallese, J.M., Kayal M., "Characteristic Parameters Evaluation of Hall Cells with High Performance", *MIXDES 2013 Conference*, June 20-22, 2013, Gdynia, Poland.

7. Paun, M.A., Sallese, J.M., Kayal, M., "Evaluation of characteristic parameters for high performance Hall cells", *Microelectronics Journal*, Elsevier, Vol. 45, Issue 9, September 2014, pp. 1194-1201.

8. Paun, M.A., Sallese, J.M., Kayal, M., "Offset and drift analysis of the Hall Effect sensors. The geometrical parameters influence", *DJNB (Digest Journal of Nanomaterials and Biostructures)*, Vol. 7, No. 3, 2012, pp. 883-891.

9. Paun, M.A., Sallese, J.M., Kayal, M., "Geometrical Parameters Influence on the Hall Effect Sensors Offset and Drift", *PRIME 2011*, Madonna di Campiglio, Trento, Italy, July 3-7, 2011.

10. Paun, M.A., "Effect of Structure Scaling on the Offset Levels for CMOS Hall Effect Sensors", *EUROCON 2015 Conference*, September 8-11, 2015, Salamanca, Spain.

11. Paun, M.A., Udrea, F., "SOI Hall Cells Design Selection using Three-Dimensional Physical Simulations", *Journal of Magnetism and Magnetic Materials*, Elsevier, Vol. 372, December 2014, pp. 141-146.

12. Paun, M.A., "Three-dimensional simulations in optimal performance trial between two types of Hall sensors fabrication technologies", Journal of Magnetism and Magnetic Materials, Elsevier, Vol. 391, October 2015, pp. 122-128.

13. Paun, M.A., Udrea, F., "Investigation into the capabilities of Hall cells integrated in a non-fully depleted SOI CMOS technological process", Sensors and Actuators A: Physical, Vol. 242, pp. 43-49, 2016.

14. Paun, M.A., Sallese, J.M., Kayal, M., "Comparative Study on the Performance of Five Different Hall Effect Devices", *Sensors Journal,* ISSN 1424-8220, Vol. 13, Issue 2, 2013, pp. 2093-2112.

15. Paun, M.A., "Single and Multi-phase Offset Numerical Estimation for CMOS Hall Effect Devices", *IWASI* 2015 Conference, June 18-19, 2015, Gallipoli, Italy. Paun, M.A., "Three-dimensional simulations in optimal performance trial between two types of Hall sensors fabrication technologies", *Journal of Magnetism and Magnetic Materials*, Elsevier, Vol. 391, October 2015, pp. 122-128.

16. Paun, M.A., Sallese, J.M., Kayal, M., "Offset Drift Dependence of Hall Cells with their Designed Geometry", *International Journal of Electronics and Telecommunications*, Vol. 59, Issue 2, pp. 16-175, ISSN (Print) 0867-6747, July 2013.

17. Paun, M.A., "Main Parameters Characterization of Bulk CMOS Cross-like Hall Structures", *Advances in Materials Science and Engineering*, Volume 2016 (2016), Article ID 6279162, 7 pages.

18. Paun, M.A., "Hall mobility study of Hall structures in two different CMOS technological processes", *EUROCON 2017 Conference*, July 6-8, 2017, Ohrid, Macedonia.

19. Paun, M.A., "On the modelisation of the main characteristics of SOI Hall cells by three-dimensional physical simulations", MIXDES 2015, June 25-27, 2015, Torun, Poland.

20. Paun, M.A., Sallese, J.M., Kayal, M., "A Circuit Model for CMOS Hall Cells Performance Evaluation including Temperature Effects", Advances in Condensed Matter Physics, Article Number: 968647, 2013.

21. Paun, M.A., Sallese, J.M., Kayal, M., "Temperature Influence investigation on Hall effect Sensors Performance Using a Lumped Circuit Model", 11th IEEE Sensors Conference, October 28-31, 2012, Taipei, Taiwan, *IEEE SENSORS Proceedings, Book Series: IEEE Sensors*, pp. 382-385, 2012.

22. Paun, M.A., "Multi-Analysis and Modeling of Asymmetry Offset for Hall Effect Structures", *Journal of Magnetism and Magnetic Materials*, Elsevier, Vol. 426, March 2017, pp. 245-251.

My presentation has now reached the end. It was my great pleasure to teach you about the Hall Effect sensors and guide you through the analysis, optimization, integration and evaluation of these magnetic sensors. I have provided you with original results which were obtained during 7 years of experience working in the domain. Two different technological processes (regular bulk and SOI) have been used in Hall cells fabrication. An overview of geometry optimization in order to satisfy strict project specifications, in terms of their offset and its temperature drift, was provided with extensive experimental results for their main parameters. Concrete dimensions and design procedures which will allow for the highest performance have been provided by the author. Circuit and physical models were developed. The growing market for automotive industry allows for the continuous development of these sensors. I thank you for your kind attention and I am looking forward to your questions.

CHAPTER 07

Low Power SoC from Bulk to FD-SOI

Pascal Urard

System Design Director
at ST Microelectronic

This talk is about Low Power SoC from Bulk to FD-SOI.

It covers the fundamentals of system integration requirements as well as technology leakages and limitations.

To get to a part explaining how to leverage FD-SOI properties for application ranging from digital to RF and finally radiation hardened circuits.

1   Content

- What IoT nodes have in common
- Low-Power Design Techniques: Pro & Cost
- Bulk CMOS Leakage Mechanisms
- Where FDSOI makes the difference?
- Body Bias in FDSOI
- Silicon results
- Analog performance & Body Bias
- Radhard & FDSOI
- Conclusion

WHAT IOT NODES HAVE IN COMMON

2   The boom of connected devices

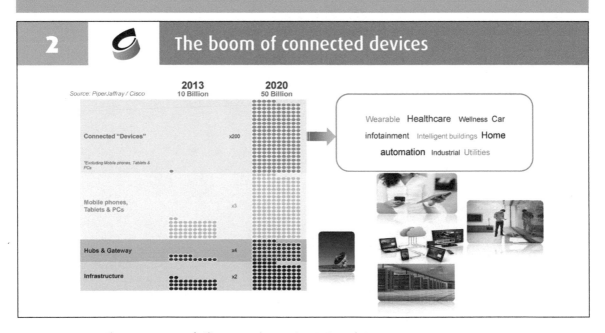

We now see the new wave of silicon needs, serving IoT markets.
This wave has started and according to specialists, should be the next driver for silicon market growth.

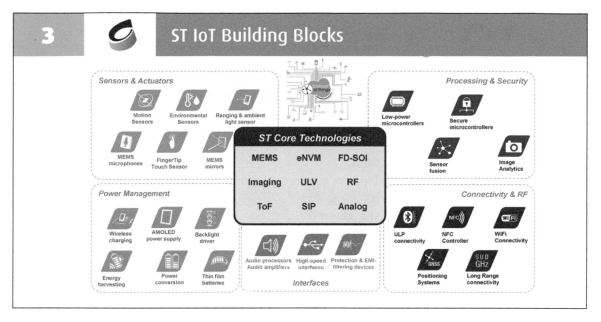

S^T is serving the IoT market with numerous building blocks, chips, SoC, enabling complete solutions.

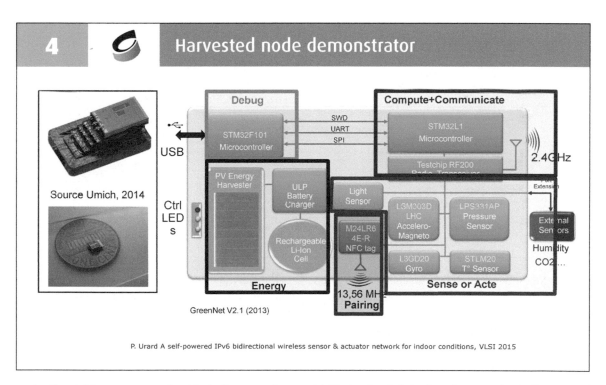

On the right you can see the block diagram of an Energy-Harvested IPv6 IoT node. This is an ST-internal project terminated in 2014 and published at VLSI 2015. On the left is the pictures of a self-powered ULP node from University of Michigan with a tiny form-factor, published in 2014.

4

As a generalization, we can consider IoT nodes are containing similar entities:

- an ULP microcontroller unit to run functional code,
- a radio,
- One or a set of sensors and/or actuators,
- An energy storage & management function: this can vary from a simple disposable battery to an energy harvester in the case of self-powered nodes, plus a power management IC to regulate

the harvester and keep the battery level within specified limits.

- Optionally a node could also contain
  - another radio. In the case of this project, we decided to include a self-powered NFC tag to ease the pairing.
  - Some debugging features enabling to monitor in-situ main ULP MCU during operation.

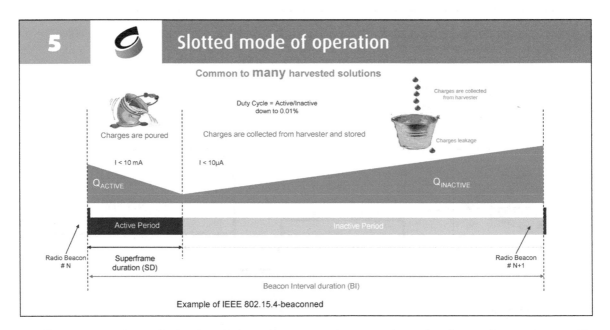

5 **Slotted mode of operation**

Common to **many** harvested solutions

Duty Cycle = Active/Inactive down to 0.01%

Charges are poured

Charges are collected from harvester and stored

Charges are collected from harvester

Charges leakage

$I < 10\ mA$

$I < 10\mu A$

Q_{ACTIVE}

$Q_{INACTIVE}$

Active Period

Inactive Period

Radio Beacon # N

Superframe duration (SD)

Radio Beacon # N+1

Beacon Interval duration (BI)

Example of IEEE 802.15.4-beaconned

Different network standard solutions have been published in the 8 last years enabling energy harvesting nodes.

The vast majority of those standards enable to have few energy harvested nodes per network, considered as end-devices, within a non-harvested network: network routers are not harvested. In those networks, each time one of the harvest devices needs to communicate, the message is received by some always-on powered node (router).

Demonstrating a fully harvested network a more tricky.

The network need to be synchronized to enable efficient communication between nodes, including routers: during pre-defined periods.

One solution is to fix one-to-one rendezvous: e.g. Standard IEEE 802.15.4e.

Another solution is to enable local synchronization using beacons (IEEE 802.15.1 or IEEE 802.15.4 beacon-enabled)

Whatever the chosen network solution, the key features of a self-powered network is in the slotted

5

mode of operation. It is composed of 2 phases.
- Phase1: The node is active during a short time period. During this phase the full system can be active: main MCU, radio, sensors/actuator.
- Phase2: The local node is mainly off, as much as possible in sleep mode (retention or equivalent), but still harvesting energy.

In harsh conditions, this phase is supposed to stay much longer than phase1 in order to balance energy harvesting and energy consumption. During phase2, only few functions remain active: ultra-low-power oscillator (RTC) to wake-up on time for next beacon, power-on reset and eventually burn-out reset to guarantee the consistency of the network data (avoid data corruption because of too low voltage on the battery), battery management (to harvest as much energy as possible but keep the battery voltage in a safe range to avoiding deep cycling and fast battery aging).

This dual phases mode is reproduced the same way continuously. Each phase is programmable at network level, in order to adapt to the available amount of energy available, depending on a given environment and/or application.

Example in case of an harvested network using standard IEEE 802.15.4 beacon-enabled: the complete sequence duration is programmable from few ms to as long as 4min 20seconds.

The active period has to be relatively short : depending on the available energy source.

It usually varies from few ms to few 10's of ms to enable self-powering.

A durable solution is found when the amount of energy harvested can balance efficiently the energy spent during active + inactive periods.

6 **T°capture + processing example**

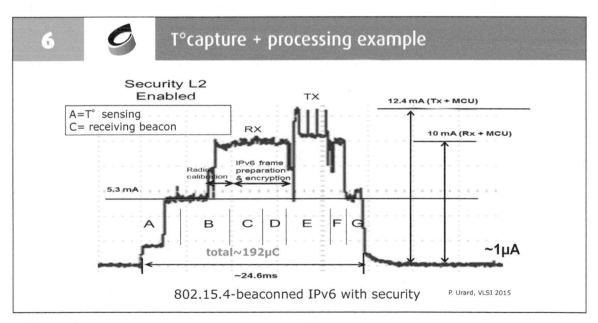

This graph represents the current consumption from the battery (~3V) for a typical 0.1°C precision temperature measurement and sending with respect to the IEEE 802.15.4 beacon-enable standard. It includes the full system: microcontroller, radio, sensors and power management.

A
- The active system wakes-up.
- MCU GPIO reconfiguration to turn them to active mode
- Charge on-board capacitance prior to turn on the radio chip

6

B

- Application : analog temperature sensing + some signal processing to ensure 0.1°C precision
- Radio: switch on the radio chip and perform radio calibration

C

- Radio : listen to channel waiting for beacon then receive & treat information included in the beacon
- Firmware: prepare IPv6 secured frame. Note: we use long Ipv6 addresses in this example.

D

- Radio: CSMA/CA (listen before talk). Can vary in length depending on the activity on the channel.
- Firmware: perform encryption

E

- send IPv6 frame

F

- receive low level acknowledgement from other sensor (parent)

G

- Switch the system back to deep sleep mode

(STOP2: all MCU registers and SRAM are in retention mode to reduce power)
- Program low speed RTC (32KHz) for next wake-up
- Reconfigure GPIOs to low power mode.

The average current consumption of the system performing temperature sensing is 5.3mA.

The average current consumption of the system in reception mode (Rx) is 10mA.

The average current consumption of the system in emission mode (Tx at 0dBm) is 12.4mA.

In this example, the total charges needed during this active period is estimated to be 192µC over 24.6ms. Average current is 7.8mA for the active period. The overall current during active+inactive periods depends on inactive period duration, and can reach as low as 1.8µA (mean value) using IEEE 802.15.4 beacon-enabled standard with the longest inactive period.

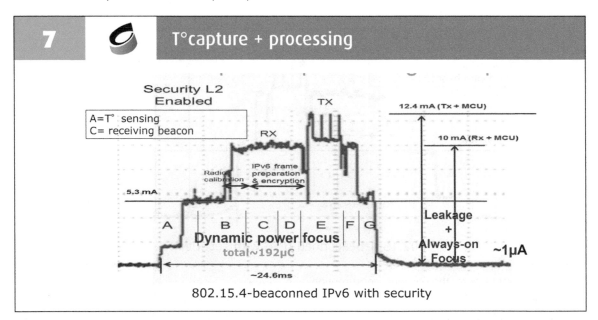

7 T°capture + processing

802.15.4-beaconned IPv6 with security

We can recognize here both active period (in green) and inactive period (in yellow).
- During inactive period, the focus of the developer is to reduce leakage and the remaining amount

of dynamic power spent in always-on blocks to the minimum.
- During active period, the focus is to minimize dynamic power.

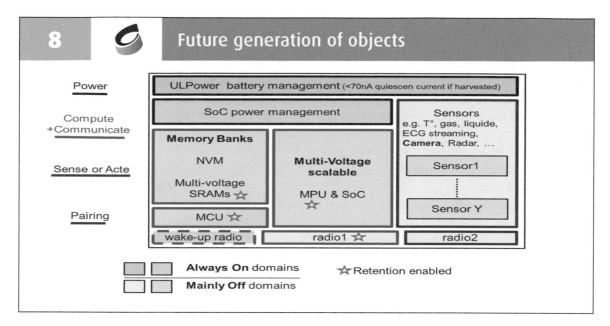

8 **Future generation of objects**

Power	
Compute +Communicate	
Sense or Acte	
Pairing	

ULPower battery management (<70nA quiescen current if harvested)

SoC power management

Sensors
e.g. T°, gas, liquide, ECG streaming, **Camera**, Radar, ...

Memory Banks

NVM

Multi-voltage SRAMs ☆

Multi-Voltage scalable

MPU & SoC ☆

Sensor1

⋮

Sensor Y

MCU ☆

wake-up radio   radio1 ☆   radio2

Always On domains

Mainly Off domains

☆Retention enabled

This slide has been borrowed to an on-going ECSEL project named PRIME, focusing on technology, design and software solutions for the next generation of IoT nodes. It globally considers the same main functions shared in Always ON and Mainly OFF domains.

Always on: power management, wake-up radio (new)

Mainly off: compute & communicate, sensors & actuators, Pairing.

LP-SOC DESIGN TECHNIQUES: PRO & COST

9 **Power Gating + Retention Flip-flops**

- Goal : keep the information in the flops during **sleep** mode.

- Flip-flop with additional "Sleep" signal (high fan-out net) and two power supplies Vddo and Vddi (switchable)

- Drawback: dual power supply in the blocks. More complex Power Mgnt

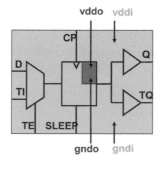

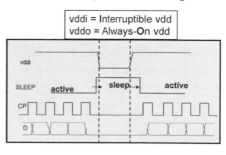

9

The basic power saving is obtain by gating the clock of digital blocks to suppress dynamic power.

This is called clock gating. Only remain leakage power.

The next step is to switch off the power in parts of the design when the clock is gated This enables to kill the leakage power.

This is called power gating.

In some cases, the information of the digital flops contained in power gated blocks has to be retained in so-called retention flops: dual powered flops where a part is always on (Vddo) when and another part is interruptible (Vddi).

The usage of retention flops enable to drastically reduce the leakage power of a given block, by several orders of magnitude.

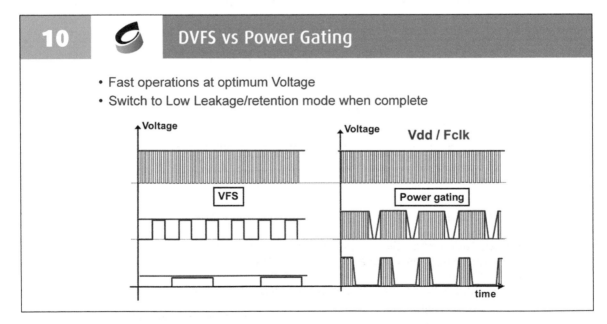

10 **DVFS vs Power Gating**

- Fast operations at optimum Voltage
- Switch to Low Leakage/retention mode when complete

VFS (Voltage & Frequency Scaling) enable one further step in power efficiency, adapting voltage to the relevant value to reach a given frequency.

In this case, leakage power will be reduced but

may still be significant.

Another technique is to switch blocks on and off, enabling to kill the leakage power between each burst of operation.

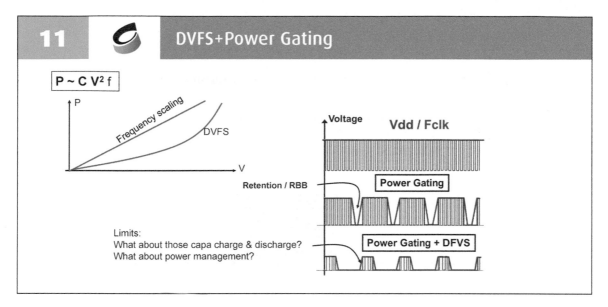

Depending on the application, the technology and the design, one of those techniques may show some benefit versus the other one.

In some case, we can also imagine mixing both of them to adapt to multiple usages.

The power has to be computed taking into account charge & discharge of design intrinsic capacitance.

In general DFVS is more energy efficient (depends on $V^2.f$) than just frequency scaling (linear scaling in power), as show in the graph.

However the power management solution at system level, can heavily impact this local power estimation, as we will see in the next slide.

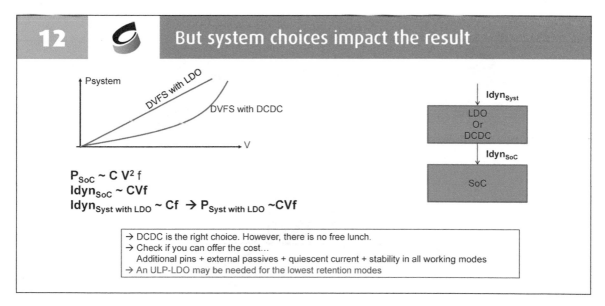

The usage of a low dropout regulator (LDO) at system level will degrade the energy efficiency of the DVFS blocks it powers.

To maintain the energy efficiency at its optimal level, the usage of a DCDC converter is needed.

The drawback of the DCDC is the complexity, and the stability in the various start-up modes it has to ensure.

In some extreme cases, it could be required to include an LDO for retention modes, on top of a DCDC for energy efficiency in functional modes, adding extra complexity to the system.

BULK CMOS LEAKAGE MECHANISMS

In the following slides, we are doing an overview of the leakage mecanisms at bulk transistor level.

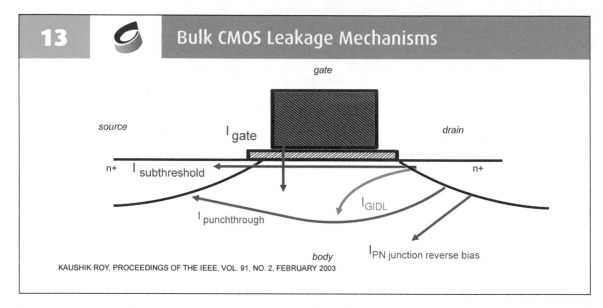

This is a simplified representation of the various leakage mechanisms of a bulk transistor.

In the following slides, we will detail each of those leakage components, and list various solutions to overcome these issues.

14 — PN Junction Reverse-Bias Current

• Leakage caused by minority carrier diffusion and electron-hole pair generation due to depletion region. Band-to-band tunnelling (BTBT) current becomes dominant for heavily doped source/drain.

The leakage is higher with supply voltage VDD increasing and Abrupt high doping profile.

Solutions could be applied at process level as well as design level.

On one hand from the process side with Well engineering (Retrograde and Halo doping), SOI.

On the other hand design could be limited to some Biasing points.

15   Subthreshold Leakage

- **Leakage caused by carriers diffusion when gate voltage is below Vth (weak inversion)..**
  This is the most important leakage in the device up to 65nm (ie: 65nm or more).

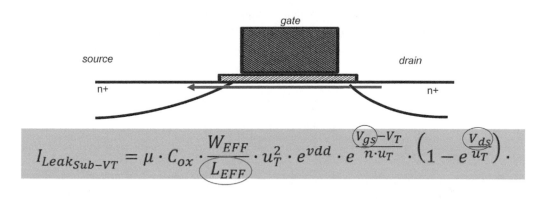

$$I_{Leak_{Sub-VT}} = \mu \cdot C_{ox} \cdot \frac{W_{EFF}}{L_{EFF}} \cdot u_T^2 \cdot e^{vdd} \cdot e^{\frac{V_{gs}-V_T}{n \cdot u_T}} \cdot \left(1 - e^{\frac{V_{ds}}{u_T}}\right) \cdot$$

In technologies larger or equal to 65nm this was the most important source of leakage.

Higher leakage with :

V_G increase in long channel devices : exponential
V_{DS} increase in short channel devices
Linear increase with Temperature increase
Solutions could be applied at process level as well as design level.

On one hand from the process side with Higher doping, Shallow junction depths, gate thicker oxide, well engineering (Retrograde and Halo doping).

On the other hand design could enlarging gate length, stacking devices, use multiple threshold devices, limit design to some Biasing points.

16   Tunneling into and through gate oxide

- **Leakage caused by oxide thickness and high electric field. Two electron tunneling mechanisms : Fowler-Nordheim and Direct tunnelling**

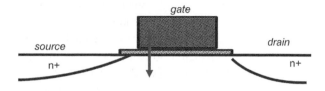

Dominant leakage in 45nm and beyond

16

Tunnelling into and through gate oxide is the dominant leakage in advanced node (45nm and beyond).

This leakage is higher when the gate voltage increase, or the technology is made with a thinner oxide.

Solutions could be applied at process level as well as design level.

On one hand from the process side with Oxide engineering.

On the other hand design could use Multiple oxide CMOSþ and limit design to some Biasing points.

---

17 **Gate-Induced Drain Leakage (I_{GIDL})**

- **Leakage caused by high field effect in the drain junction due to narrowing of the depletion layer near the surface, generating carriers into the substrate and drain**

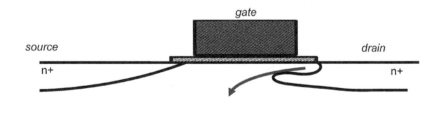

Gate-Induced Drain Leakage is caused by high field effect in the drain junction due to narrowing of the depletion layer near the surface.

It generate carriers into the substrate and drain. It increase exponentially when VG and VDS difference increase.

Other contributions such as thin Oxide and abrupt and strong doping profile could apply.

Solutions could be applied at process level as well as design level.

On one hand from the process side with Oxide engineering, drain doping, interface engineering (substrate/insulator).

On the other hand design could put the function in it's sleep-mode and disable the transistor.

18   Punchthrough leakage

- Leakage caused by merging of the depletion regions, combined with high drain voltage, allowing carriers to go through them.

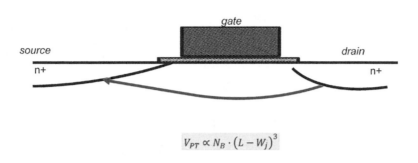

$$V_{PT} \propto N_B \cdot (L - W_j)^3$$

The Punch through Leakage is caused by merging of the depletion regions, combined with high drain voltage.

It allows carriers to go through them. When supply voltage VDD increase above Punch voltage "VPT", more carriers goes through.

When channel length decrease it's another contributor for a higher leakage.

Solutions could be applied at process level as well as design level.

On one hand from the process side with Additional implants, retrograde and halo doping.

On the other hand design could use non minimum gate length and adapt biasing points.

19   Libraries mixed-Vt

- Timing / leakage tradeoff : HVT = better leakage, lower speed
  LVT = better speed, higher leakage
- No area penalty : same layout & abstract
- Implementation compatible with std flow (easy)

Vt in 65nm design	Power (mA)		Frequency	Cell distribution		
	Dyn	Leak-125c		LVT	SVT	HVT
LVT	114	7X	131	100%		
SVT	Base100	1X	base100		100%	
HVT	97	0.1X	60			100%
Mix LVT/SVT	103	1.9X	133	15%	85%	
Mix SVT/HVT	99	0.3X	102		26%	74%

The above figures are Silicon data

Dynamic power varies only a little when changing the Vt, but leakage power is heavily impacted, as well as performance of a design.

This table shows in the 2 lower green lines, the possible tradeoffs of mixed Vt design solutions and dynamic/leakage power balancing.

The challenge for the designer is to optimize power consumption for the minimal impact in terms of performances.

The Vt mixing is offers the most efficient trade-offs.

With modern tools and flows, this tradeoff is completely automated in design flow and mixed Vt libraries.

20 PolyBias – or LargeL libraries

- Goal : reduce leakage by enlarging transistor L
- Pro
  - Less impact on delays than higher Vt : can still go down in voltage
- Cons
  - Area : std-cell foot-print can be larger. Or reduction of Contact-Poly distance --> tricky DRM

'L' increase	+23% (Vt$_1$)	+23% (Vt$_2$)	+7.7%
Leakage reduction	0.3X	0.3X	0.6X
Area penalty at block level	<5%	<5%	<1%
Delay penalty	+30-35%	+20-25%	+10-15%
	New libs with a higher poly pitch		Same lib, various PB

Last but not least, enlarging the L value of the transistors in the standard cells enable to save significant power at block level: x3 for a+5% area penalty.

Library providers have introduced fine-grain optimization of this principle by reducing poly-contact distance to the minimum inside the standard-cells, in order to enlarge the L value of the transistors as much as possible, keep the poly-pitch unchanged.

This added complexity to the design-rules manuals, but enabled to save significant power for a little area penalty (last column on the right). This method is called poly-biasing (PB).

21 Why industry is reluctant to SubVt SoC?

Local variability*: Avt imposes a limit to (Vdd-Vt)
- for a good yield at t0
- for a better reliability with silicon aging

*On top of SS/TT/FF corners which represent global variability

Avt impacts the saturation current in the law :

$$I_{sat} = f(V_{gs}-Vt_{effective})^2$$
$$= f(V_{gs}-Vt_{nom} +/- n\, Avt/sqrt(2*W*L)\ )^2$$

Distance Vdd to Vt

Size of xtor (design)

Local variability in mV.µm

Distribution of I$_{sat}$ current in similar transistors of a given chip

Avt=1

Avt=4

I$_{sat}$

Weaker xtor of the chip

Weak I$_{sat}$

Strong I$_{sat}$

The limit of Vdd reduction is fixed by the increase of the local variability at low Vdd: the local variability increases quadratically as soon as Vdd is close enough to Vt (depends on the size of the transistor). This variability is called local because it comes on top of global SS/TT/FF corners.

21

The local variability is modelled by the parameter AVt in the device model. It is expressed in mV.µm.

The factor 'n' in the equation is linked to the number of occurrences in the design, and can be estimated through statistical calculation. On the graph on the right, we see the impact of an AVt=1mV.µm versus AVt=4mV.µm on a given design: statistically, the worst transistor of a design could come out with quite no more current on a variable process at low Vdd.

The more variability, the more timing margins, leading to lose performance and increase power.

This quickly balance the advantage of reducing Vdd.

This is one of the main reason why large SoC designers are reluctant to adopt SubVt design solutions.

WHERE 28FD-SOI MAKES THE DIFFERENCE?

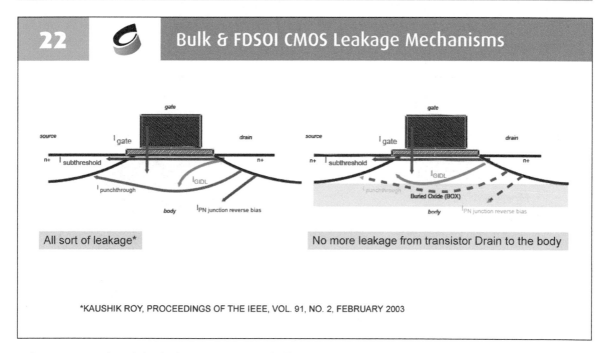

22   **Bulk & FDSOI CMOS Leakage Mechanisms**

All sort of leakage*

No more leakage from transistor Drain to the body

*KAUSHIK ROY, PROCEEDINGS OF THE IEEE, VOL. 91, NO. 2, FEBRUARY 2003

This is a reminder of the leakage current in a bulk CMOS transistor is proposed and it can be compared with FDSOI remaining leakage components.

The main advantage in terms of leakage is that there is no more leakage from transistor Drain to the body.

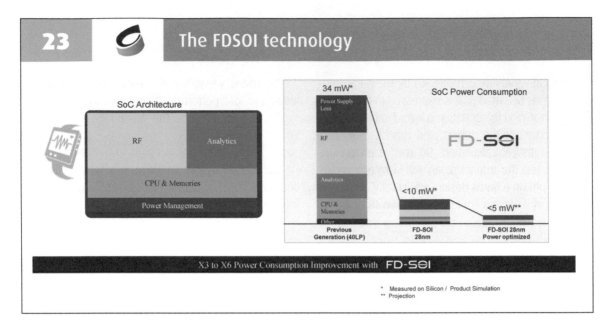

The FDSOI technology is enabling to drastically optimize power/performance of ULP designs, especially in the case multi-modes of operations.

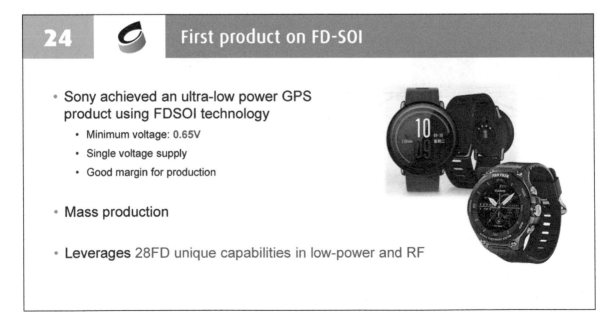

Example of industrial realization.

Press releases showing FD-SOI adoption.

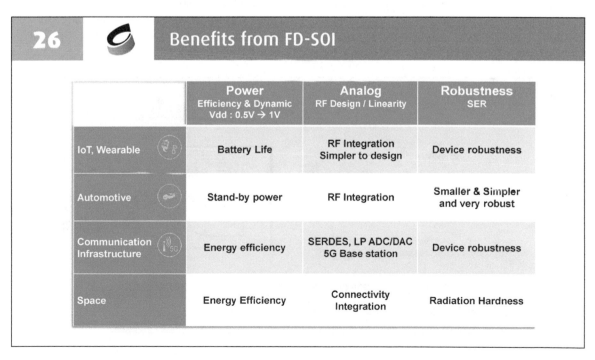

This slide describes the main benefit we can list from on Power tradeoff, Analog designs or Robustness to Soft Errors, for each market segment: IoT & Wearable, Automotive, Digital communications, and Space Applications.

As an example: Automotive radars are currently based on 24GHz devices. 28nm FD-SOI supports the next move to 77GHz with unmatched performances /cost / integration trade-offs

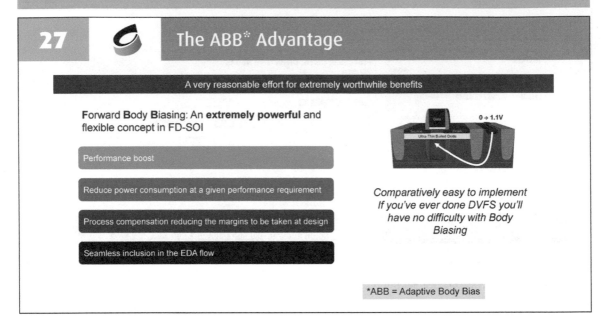

This slide shows the main advantages of Adaptive Body Bias (ABB).

ABB is using the back side gate to adjust the Vt value of the transistor to most appropriate value depending on the application.

The 3 main advantages can be targeted and balanced/mixed depending on the application needs.

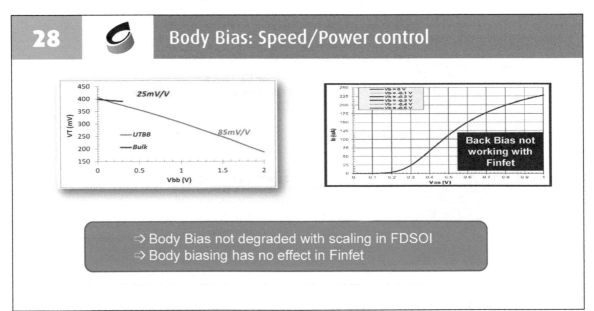

In FDSOI, back-biasing consists of applying a voltage just under the BOX of target transistors. Doing so changes the electrostatic control of the transistors and shifts their threshold voltage VT, to either get more drive current (hence higher performance) at the expense of increased leakage current (forward back-bias, FBB), or cut leakage current at the expense of reduced performance (reverse back-bias, RBB). The left figure illustrates the concept of FBB for NMOS.

- For NMOS, FBB is when Vbs>0, that is bringing gnds above gnd – and reverse bias is the opposite.
- Conversely for PMOS, FBB means Vbs<0, that is bringing Vdds below Vdd.

Back-biasing can be utilized in a dynamic way, on a block-by-block basis. It can be used to boost performance during a limited periods of time when maximum peak performance is required from a block. It can also be used to cut leakage during some periods of time when limited performance is not an issue (one obvious example being when blocks are in stand-by).

In summary, back-bias offers a new and efficient knob on the speed/power trade-off.

On the right graph, all the curves are superposed whatever the bias voltage value (Vb).

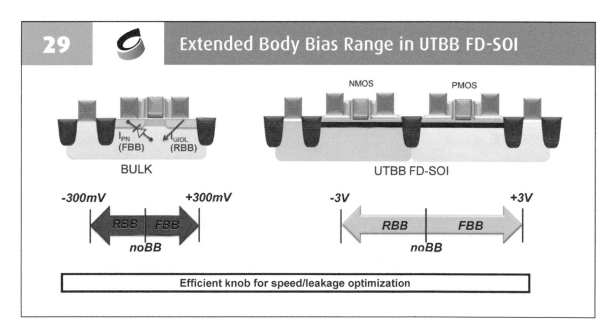

The maximum amplitude of the bias, and therefore its impact, is rather limited with bulk CMOS: the conventional bulk transistor structure does not allow reaching outside a +-300mV body bias range, otherwise leakage currents become unacceptable. With planar FD, because the buried oxide provides complete dielectric isolation of the source and drain, the possible back-bias range is much wider and large performance boost factors can be obtained. The limiting factor is now the p-n junction between wells, which must not be forward biased e.g., by making the p-well bias significantly higher than the n-well bias.

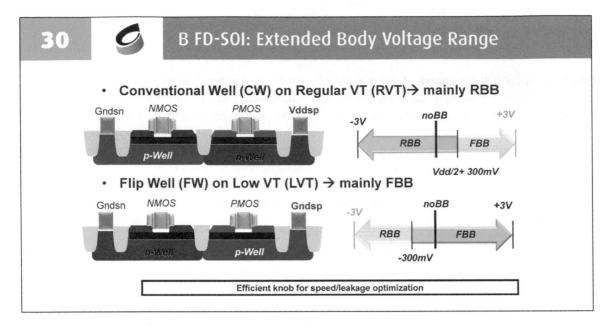

This is the illustration of 28FDSOI Body Bias trade-off.

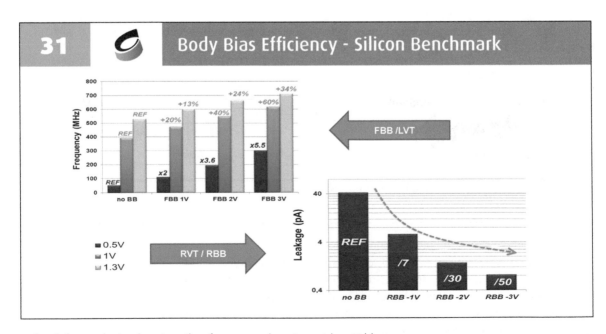

The left graph is showing the frequency boost using FBB on LVT transistors, depending Vdd value (color) and FBB value. FBB impact is more efficient at low Vdd.

The right graphs is showing the leakage reduction when applying RBB on RVT transistors

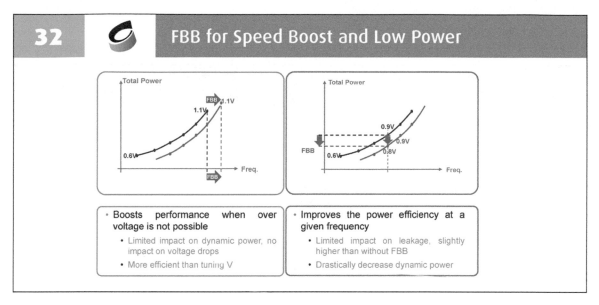

These graphs display two kinds of trade-off using FBB in a give condition on a design:

- the benefit in terms of frequency at constant power
- or the benefit in terms of power at constant frequency

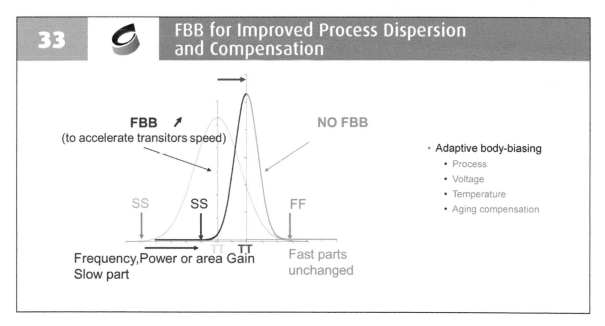

This graph displays the benefit of the compensation of process dispersion, using FBB.
   The slower the part, the more FBB is applied.
- Fast parts are not affected (FBB=0)
- Standard process parts (TT) are speed-up a little
- Slow parts are speed-up frankly with a higher FBB value

SILICON RESULTS

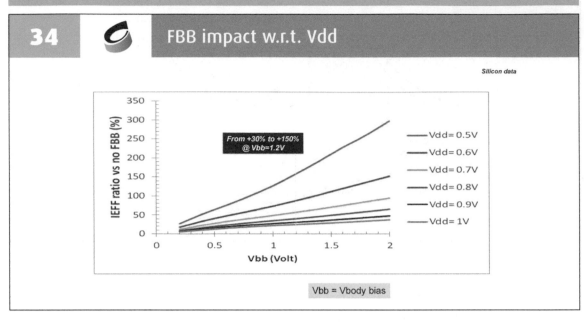

34 FBB impact w.r.t. Vdd

Silicon data

From +30% to +150%
@ Vbb=1.2V

Vbb = Vbody bias

S ilicon measurement.
    Transistors effective current (Ieff) improvement, depending on Vdd and FBB (Vbb) values.

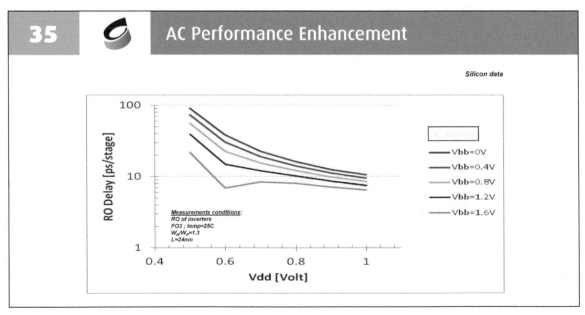

35 AC Performance Enhancement

Silicon data

Measurements conditions:
RO of inverters
FO3 ; temp=25C
W_N/W_P=1.3
L=24nm

S ilicon measurement.
    This graph represents a ring-oscillator delay impact (representative of 1/Frequency) depending on Vdd and FBB (Vbb) values.
    FBB is more efficient in percentage at low voltage.

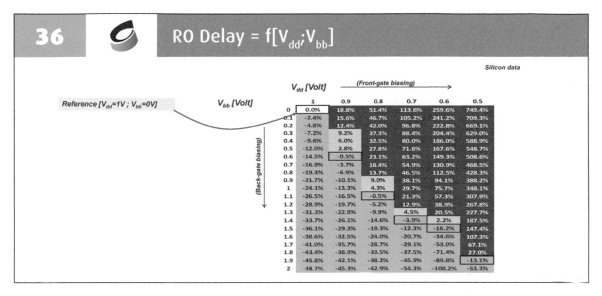

E xtracted from the two previous slides, we can build this matrix with Vdd and Vbb entries.

The reference is the white cell in the upper left corner: Vdd=1V, Vbb=0

Depending on Vdd and Vbb values:

- **In red,** performance is degraded
- **In green,** performance is better

The diagonal cells shows couples (Vdd, Vbb) with no/little degradation of the performance, but with a much better energy efficiency than the reference solution in the upper left corner.

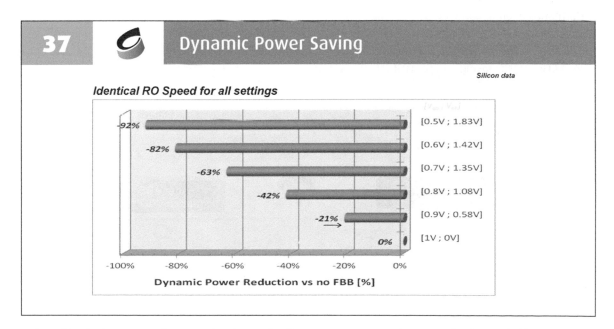

T his table displays in another way the diagonal cells of the previous slide. It directly shows the power savings versus the original solution (Vdd=1V, Vbb=0), for various (Vdd, Vbb) couples.

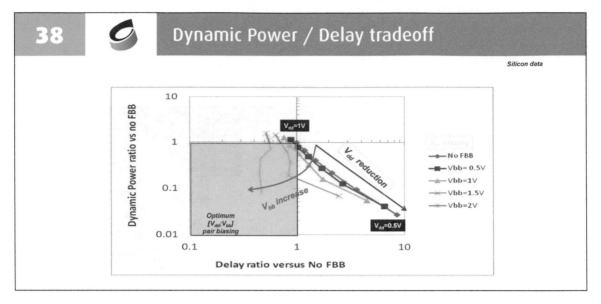

The central point is the reference: no body bias, Vdd=1V.

The vertical axis represent the dynamic power ratio versus the reference.

The horizontal axis represent the delay ratio (performance) versus the reference.

The desired quadrant is the green one : no performance degradation with power reduction.

The blue curve represents a reduction Vdd with body bias: power saving at the cost of important performances degradation.

We seen that with 1.5V of FBB, we can have point in the green quadrant. Each point of the purpul curve shows various values of Vdd.

With 2V of FBB, the orange curve demonstrates both power saving and delay improvement at the same time

Conclusion: when we introduce enough FBB, we can more easily find solutions in the green quadrant.

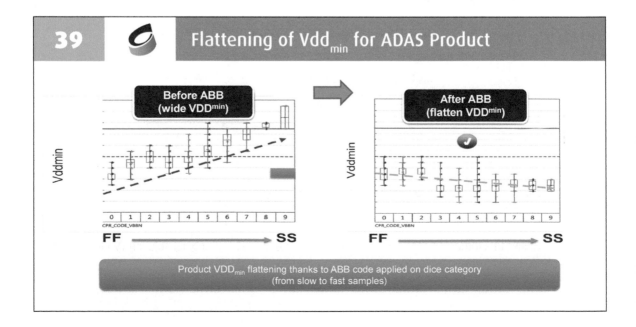

39

Silicon measurements of a SoC targeting ADAS market

These graphs show the product dispersion regarding Vddmin value, depending on process corner.

LEFT: before ABB, only fast parts are passing the Vddmin target (dash line)

RIGHT: after ABB, all the production corners are within Vddmin specification. Only few % of the production does not meet the target (touching the dash line).

---

40 **Performance/Reliability Optimized with ABB**

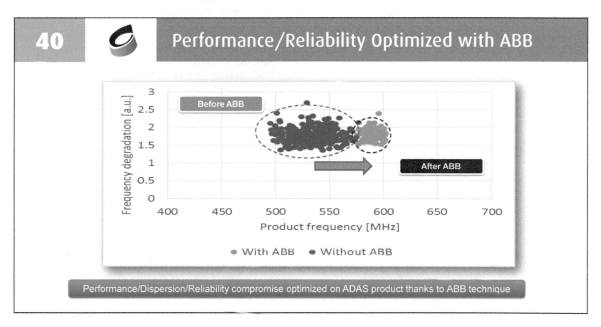

Performance/Dispersion/Reliability compromise optimized on ADAS product thanks to ABB technique

---

The process corners dispersion is reduced thanks to ABB, taking into account aging margins.

Please note there is no performance degradation due to aging : this is a pure horizontal shift only.

ANALOG PERFORMANCE AND BODY BIASING

41 **Body Biasing & Analog Blocks Area Reduction**

Birds view of FBB benefits for digital and analog parts.

FBB positively impacts gm, gd, Ft by reducing Vt value.

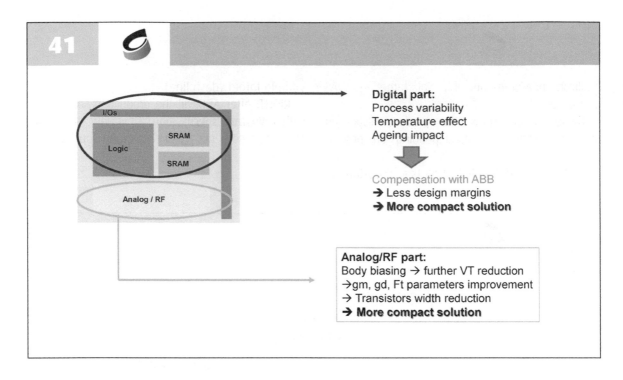

41

Digital part:
Process variability
Temperature effect
Ageing impact

Compensation with ABB
→ Less design margins
→ **More compact solution**

Analog/RF part:
Body biasing → further VT reduction
→gm, gd, Ft parameters improvement
→ Transistors width reduction
→ **More compact solution**

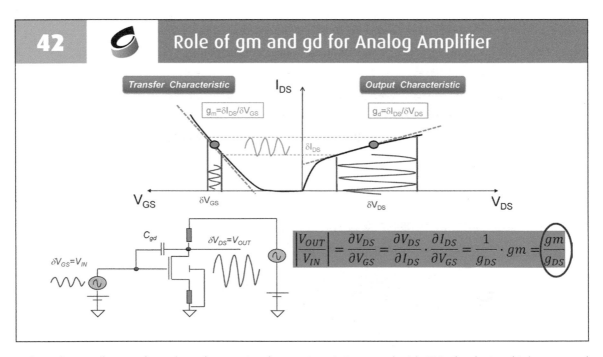

42 **Role of gm and gd for Analog Amplifier**

These figures illustrate how the voltage gain of a transistor is improved with FBB: thanks to a higher gm and lower gd.

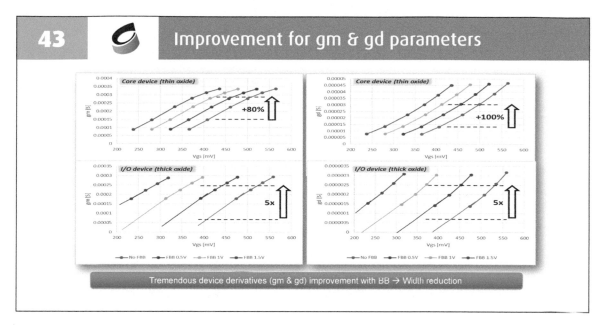

43 Improvement for gm & gd parameters

Tremendous device derivatives (gm & gd) improvement with BB → Width reduction

These tables show the FBB impact on gm and gd parameters.

Up to a factor x2 can be achieved depending conditions for core devices (thin oxide), and up to x5 improvement can be achieved for I/O devices (thick oxide).

Upper graphs : core devices
Lower graphs : I/O devices

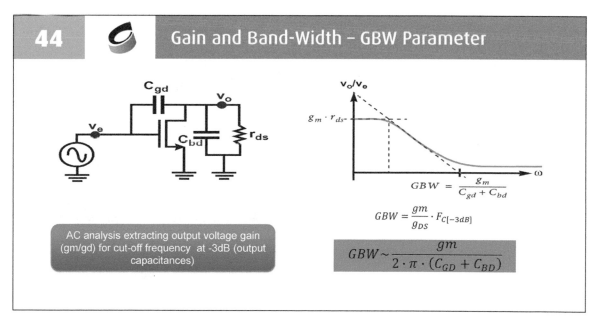

44 Gain and Band-Width – GBW Parameter

AC analysis extracting output voltage gain (gm/gd) for cut-off frequency at -3dB (output capacitances)

$$GBW = \frac{g_m}{C_{gd} + C_{bd}}$$

$$GBW = \frac{gm}{g_{DS}} \cdot F_{C[-3dB]}$$

$$GBW \sim \frac{gm}{2 \cdot \pi \cdot (C_{GD} + C_{BD})}$$

This is a reminder of the definition of the GBW (gain x bandwidth product).

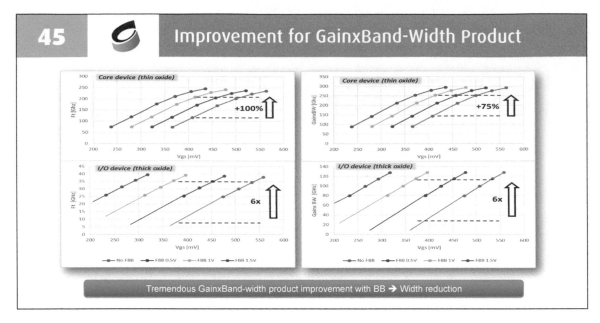

These table show GBW product (Gain x BandWidth product) improvement, depending on FBB values: up to x2 can be achieved depending conditions for core devices (thin oxide), and up to x6 improvement

can be achieved for I/O devices (thick oxide).

Upper graphs : core devices
Lower graphs : I/O devices

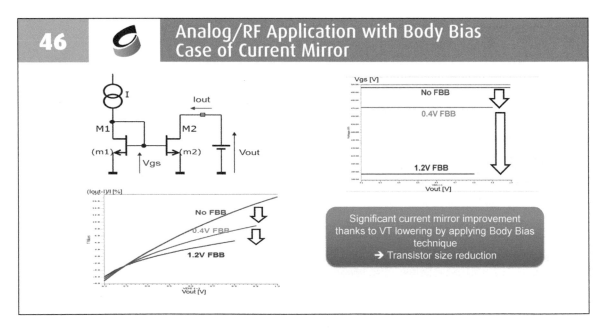

These graphs show the performance of an NMOS current mirror, depending on Vgs values. The current mirror quality (Iout-I)/I is enhanced when increasing FBB (left graph).

Two possible tradeoff directions at constant mirror quality:

• Vgs can be reduced when applying FBB (right graph)
• Transistors size can be reduced when applying FBB.

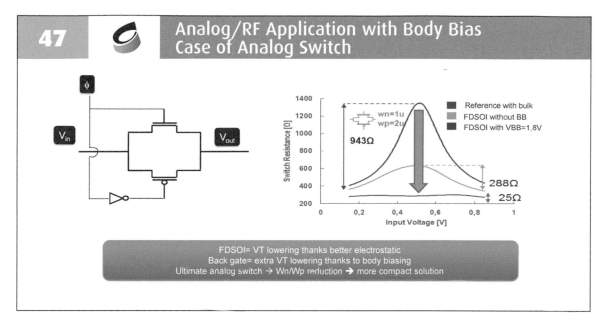

47 Analog/RF Application with Body Bias
Case of Analog Switch

FDSOI= VT lowering thanks better electrostatic
Back gate= extra VT lowering thanks to body biasing
Ultimate analog switch → Wn/Wp reduction → more compact solution

Numerical example of an analog switch (pass-gate) resistance reduction using FBB.
R = f (Vout-Vin)

The resistance R can be reduced by a factor x11 with Vbb=1.8V, compared to Vbb=0.

RAD-HARD AND FDSOI

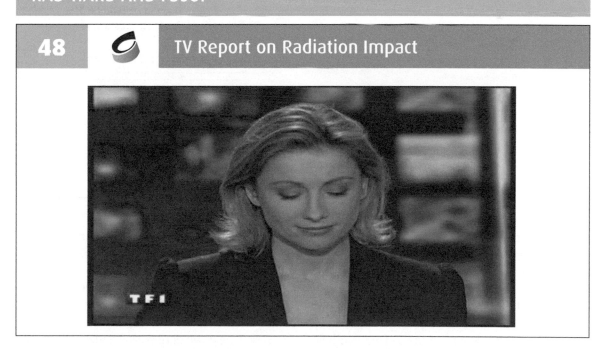

48 TV Report on Radiation Impact

As a foreword, TV Report on Radiation Impact on ICs, Broadcasted on the French National News.

Discussing the threat of cosmic particle's to which electronics is sensitive.

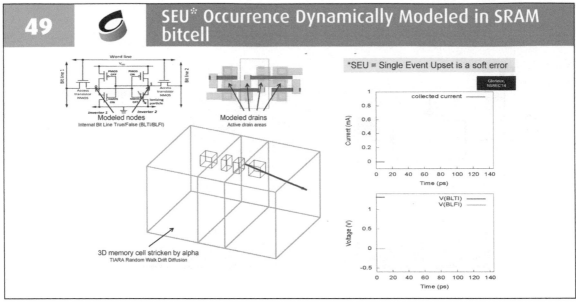

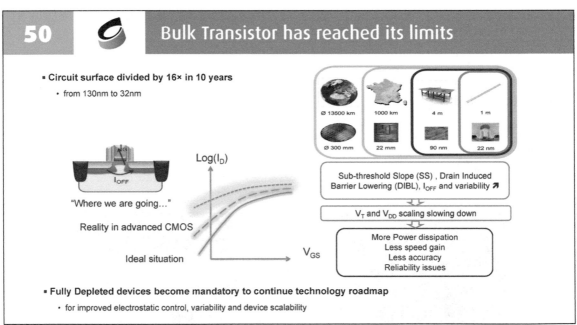

As you know, the deca-nanometer technologies face lot of issues due to the extremely small spacing between the drain and de source, which jeopardizes the good gate to channel electrostatic control

This leads to:

- An increase of the leakage current (far away from the gate) degrading the SS (sub-threshold slope).
- A shift of the VT due to SCE and DIBL degrading the IEFF (speed)
- An increase of the VT fluctuation due to a higher sensitivity to process variability
- It results in:
- A slowing down of VT and VDD scaling
- Power dissipation issues affecting the reliability of the devices
- A slowing down of Lg scaling trading off power/speed/variability

51   Outstanding intrinsic radiation robustness in FDSOI

- **Very small volume of charge collection**
  - ion-induced charges only collected from Si film
  - not below Buried Oxide (BOX), not beyond STI

- **Best radiation isolation vs. other technologies**
  - 160/70× smaller Si film than PDSOI 130/65nm
  - no multiple charge collections as for FinFET
  - Latchup immunity by construct

- **Very low gain of parasitic bipolar**
  - parasitic BJT revealed by radiation
  - $Q_{collected} = \beta \times Q_{deposited}$
  - $\beta < 15$ in PDSOI vs. $\beta < 3$ in FDSOI

ST papers, Roche et al.
IEDM'13
NSREC'14
IRPS'14
RADECS'14
RASEDA'15

> Outstanding intrinsic radiation performances
> SER (bit flip) and SEL (Latch-up) in 28nm FDSOI

52   28nm UTBB FD-SOI: Best SER-SRAM/FF at sea-level

- **Very low neutron-SER for SRAMs <10 FIT/Mb**
  - 100× better than BULK 28nm
  - ECC-SRAM not systematically required
  - 99% single bit errors

- **Very low neutron-SER for FFs <1 FIT/MFF**
  - triplication or lockstep not systematically required
  - rad-hard FFs tested immune

- **Single Event Latchup immunity**
  - tested immune with space ions 125°C/1.3V

- **Alpha quasi-immunity <1 FIT/Mb**
  - 1000× better than in BULK 28nm
  - no need for ultra-pure alpha packaging

ST, Roche
IEDM'13
IRPS'14
NSREC'14

Experimental SRAM Failure-in-Time (FIT) test data

FD-SOI

neutron-SER in FIT/Mb

28nm bulk — 28nm FDSOI

ST 65LP | ST 45LP | Vendor A 45LP | Vendor A 40G | Vendor B PDSOI45 | ST C32 | ST C28 | Vendor A C28 | ST FDSOI28

Gain w.r.t. BULK	UTBB FDSOI	FinFET
Alpha	1000×	15×
Neutron	100×	10×
Latchup	immune	not reported
Multiple Cell Upsets	99% single bit	4+ cells

ST, Roche, CPMT, Oct'14

TSMC, Fang, CPMT, Oct'14

> Best terrestrial SEU/SEL resilience
> in FDSOI 28nm

Note : FIT = Failure-in-Time = Fails per billion-chip hours

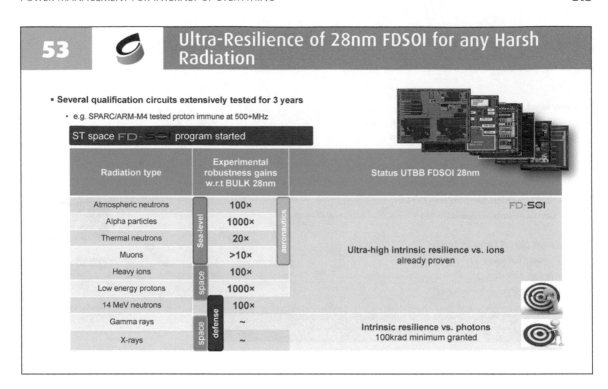

53 Ultra-Resilience of 28nm FDSOI for any Harsh Radiation

- Several qualification circuits extensively tested for 3 years
  - e.g. SPARC/ARM-M4 tested proton immune at 500+MHz

ST space FD-SOI program started

Radiation type	Experimental robustness gains w.r.t BULK 28nm		Status UTBB FDSOI 28nm
Atmospheric neutrons	Sea-level	100×	aeronautics
Alpha particles		1000×	
Thermal neutrons		20×	
Muons		>10×	Ultra-high intrinsic resilience vs. ions already proven
Heavy ions	space	100×	
Low energy protons		1000×	
14 MeV neutrons	defense	100×	
Gamma rays	space	~	Intrinsic resilience vs. photons 100krad minimum granted
X-rays	defense	~	

FD-SOI

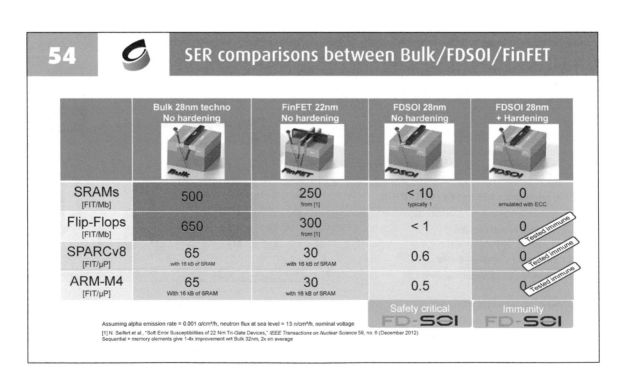

54 SER comparisons between Bulk/FDSOI/FinFET

	Bulk 28nm techno No hardening	FinFET 22nm No hardening	FDSOI 28nm No hardening	FDSOI 28nm + Hardening
SRAMs [FIT/Mb]	500	250 from [1]	< 10 typically 1	0 emulated with ECC
Flip-Flops [FIT/Mb]	650	300 from [1]	< 1	0 Tested immune
SPARCv8 [FIT/μP]	65 .with 16 kB of SRAM	30 with 16 kB of SRAM	0.6	0 Tested immune
ARM-M4 [FIT/μP]	65 With 16 kB of SRAM	30 with 16 kB of SRAM	0.5	0 Tested immune
			Safety critical FD-SOI	Immunity FD-SOI

Assuming alpha emission rate = 0.001 α/cm²/h, neutron flux at sea level = 13 n/cm²/h, nominal voltage

[1] N. Seifert et al., "Soft Error Susceptibilities of 22 Nm Tri-Gate Devices," *IEEE Transactions on Nuclear Science* 59, no. 6 (December 2012)
Sequential + memory elements give 1-4x improvement wrt Bulk 32nm, 2x on average

55 Conclusion

In all harvested nodes, energy efficiency is a central concern.
- leakage has to be managed.
- but an holistic system approach is highly recommended

28FDSOI, thanks to the FBB is the right answers for:
- **In Digital :** where dynamic power is of importance. Typical: wearable with significant digital
- **In RF :** for Low-Power RF with High-Performances
- **RadHard :** In harsh conditions or ultimate low voltage

IoT is fragmented and low-cost, older nodes still have a nice future, and are still growing in volume.

Acknowledgments :

The speaker would like to acknowledge the following people for their support and the material they shared: The Technology & Design Platform Team of STMicroelectronics Special thanks to direct contributors: Franck Arnaud, Philippe Roche, Giorgio Cesana, Christophe Bernicot, Philippe Flatresse and Sylvain Clerc.

CHAPTER 08

It's all about time

Mathieu Coustans

Ecole Polytechnique Fédérale
de Lausanne (EPFL), Switzerland

In the IoT, a large number of sensors will be installed anywhere and everywhere on the human body or in our environment. They will collect a range of physical and environmental information which will be sent to the internet via some sort of connectivity path. Current application microcontrollers including connectivity, computing capability and sensors report an average power of 100 µW. This 100 µW output of power is made up of 40 µW for the processor and 20µW for sensor conditioning and the remainder is used for the data converter and the radio link.

A conservative forecasts state that 3 billion devices will be installed by 2020, which represents 300 mW, or a third of the production capability of a nuclear plant. Intermittent operation is one of the key techniques for low power VLSI systems including radio, which suggests the use of energy-efficient Time Division Multiple Access (TDMA) protocols such as IEEE 802.11. These techniques achieve low power and a long battery life by aggressively cutting off power supply. Total average power (of the system) is mainly determined by that at sleep mode. So, we must minimize the power used during sleep mode as low as possible. A real-time clock circuit must stay awake at all times (even though other circuits are in sleep mode). Therefore, an ultra-low power clock generator for RTC is required. The presentation entitled "It's all about time" was motivated by the tremendous need for energy efficient and frequency-scalable clock generators.

The investigation presented was funded by and realized in collaboration with an industrial partner, EM-Microelectronics, part of the Swatch Group Ltd Electronic Systems.

1 Outline

- State of the art

Micro Mote M³
15nW stdby
304nW (motion detection)
180nm
Kim, G. et al, VLSI'14

SleepWalker
1.7µW stdby
7µW/MHz@0.4V(25MHz)
65nm
Bol, D. et al, JSSC'13

TI CC2650
1µA stdby
6mA (RX/TX mode)
TI, CC2650,'15

HOW TO ACHEIVE STANDBY CURRENT ?

- Architecture presentation

- Results and discussion

Saying that it's all about time may sound bold but the clocking of electrical circuits is a crucial issue since clock signals are used to establish the flow-of-time within the world of electronics.

Flow-of-time will remain an eternal issue, before, during and after the rise of silicon IC technology. IC clocking could be regarded as the fourth major IC design technology.

The body follows on from the motivation section discussing the motivation to provide a lecture related to frequency generator and time keeping in a seasonal school on **Power Management for Internet of Everything**.

- On one hand, the importance of time to battery-operated systems, suggesting the need to consider energy instead of power in the Internet of Things. Timed events or duty-cycle functions are widely used for Internet of Things.
- On the other hand, the Internet of Everything is about anything-to-anything communication and requires more than ever an understanding of the flow-of-time.

The course notes then introduce state-of-the-art techniques for clock generation and compare a crystal oscillator and RC oscillator.

The requirements of applications with the example of radio pairing suffering from time base mismatch, which is rarely avoidable, clearly suggest that if the object needs to supply the radio with 200 ms, then most of the power consumption comes from one phase which could have been avoided. Then, the most commonly-reported structures in the literature are discussed.

And this introductory section then looks at the transition requirements for a time base in the Internet of Things compared to the example of radio pairing in the paper mentioned.

The next core section system, which discusses the proposed architecture with the four key points that make the circuit and system technique robust to process variation, supply voltage and temperature variation, in other words a reference. As the function generates frequency, it is a frequency reference that embeds a temperature compensation scheme in order to be robust to temperature variations.

The previous section presented some argumentation regarding the need to use a reference that is robust to temperature and voltage variation as well as process variation. This leads us to the final section, which presents the assessment results for silicon and benchmarks these results against other publications.

2 Motivation (II)

- **How much energy out of a rain drop ?**
  - **We all felt once a rain drop on our skin, what if we could convert that energy with 100% efficiency ?**

$$K_E = \frac{1}{2} \times M \times V^2 \qquad M \sim 4.19 \times 10^{-6} Kg \qquad V \sim 8.9 \frac{m}{s}$$

$K_E \sim 166\ \mu J \rightarrow 166\ \mu W \times s$

- **How much watch cycle would that run ?**
  - **Recent crystal watch burn less than 1µA @ 1.55V (1.55µW).**
  - **Mostly in the motor**
- **107 second clock hand steps, Almost 2 minutes**

rtment of Chemical Engineering and Materials Science, University of Minnesota.
o the Atmosphere. 3rd ed. New York: McGraw Hill, 1978: 107.
ography Manual. 1974. 5th ed. N.p.: Kendall/Hunt, 2003. 127.

When talking about ultra-low power, it is not always easy to communicate the orders of magnitude which cannot be easily apprehended by everyone.

Everyone has a sense of what it means to be hit by a falling drop of water.

Kinetic energy contained in a falling raindrop may be easily derived through physics and using some scientific explorative data.

If there was a means to harvest this energy, it would represent an average of 166 µJ or 166 µW per second. This figure could be 1mJ in more optimistic cases.

Stretching this amount of energy out over time would allow, for example, to power a modern analogical quartz watch for almost 2 minutes.

The consumption of a modern quartz watch is a bit less than 1µA @ 1.55V (1.55µW).

Roughly 90% of this energy is used by the watch motor to move the hands.

When discussing Internet of Everything or Internet of Things, the activity is duty-cycled and the overall system is either battery or harvester and capacitor or super-capacitor supplied.

It's important to include the time dimension in the system design as it concerns the packet of energy supplied.

3

The concept of the Internet of Everything originated at Cisco, who defines IoE as "the intelligent connection of people, process, data and things."

Because in the Internet of Things communications are machine to machines, some say that "IoT and M2M are sometimes considered synonymous".

The more expansive IoE concept includes, besides M2M communications, machine-to-people (M2P) and technology-assisted people-to-people (P2P) interactions.

The concept of linked data was introduced by Tim Berners Lee [internet inventor, back in 1989 at EUROPEAN CENTER FOR NUCLEAR RESEARCH (CERN -

Motivation (II)

- *Internet of Everything ?*

People — Connecting People in More Relevant, Valuable Ways

Process — Delivering the Right information to the Right Person (or Machine) at the Right Time

Data — Leveraging Data into More Useful Information for Decision Making

Things — Physical Devices and Objects Connected to the Internet and Each Other for Intelligent Decision Making

- **Moving from a Machine to Machine toward anything to anyhting connectivity. Including pepole and security risks fastest growing being theift. Time stamp and legal time matter**

Switzerland).

This picture appears a bit Cisco-centered but actually it concerns two-way communication; i.e. Any-to-Any.

There is always the risk that the man in the middle was not supposed to be informed. Security and particularly banking relies on a precise legal time.

4

There are several clock generators. Among them, a crystal oscillator is widely used because it demonstrates high accuracy and its power consumption is extremely low.

However, since it needs an off-chip component, implementation volume becomes large.

On the other hand, an RC oscillator is widely used as an on-chip CMOS oscillator.

It can be fully integrated on a chip and it shows reasonable accuracy.

Accuracy has a vague definition. In case of the above publication, it has to do with the temperature

State of the Art (I)

Type	Crystal oscillator [1]	CMOS RC oscillator [2]
Implement	Off-chip XO	Fully on-chip !! CMOS LSI chip / PCB
Frequency	32.768 kHz	3 kHz
Accuracy	< ±100ppm	> ±1000 ppm
On-chip	No	Yes
Power	15 nW @ 1V	4.7 nW

On-chip CMOS RC oscillator for IoT device's RTC

[1] A. Shrivastava, "A 1.5 nW, 32.768 kHz XTAL Oscillator Operational From a 0.3 V Supply," *IEEE J. Solid-State Circuits*, vol. 51, no. 3, pp. 686–696, Mar. 2016.
[2] T. Jang, M. Choi, S. Jeong, S. Bang, D. Sylvester, and D. Blaauw, "A 4.7nW 13.8ppm/°C self-biased wakeup timer using a switched-resistor scheme," *2016 IEEE Int. Solid-State Circuits Conf.*, pp. 102–103, 2016.

deviation from 25°C to its highest range.

RC oscillator for an IoT device's RTC could be acceptable, depending on the application.

90% of the frequency control market (total value in 2016 4.5 billion dollars) remains dominated by the crystal oscillator.

5   State of the Art (II)

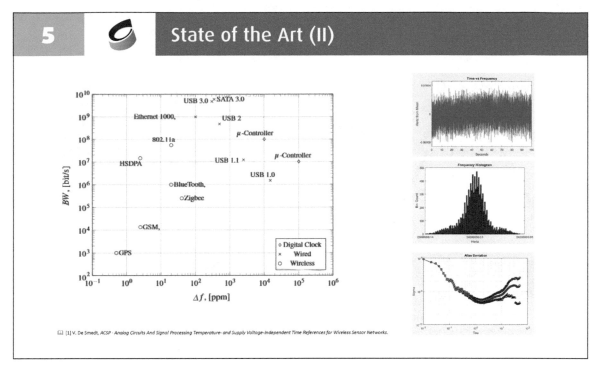

[1] V. De Smedt, ACSP · Analog Circuits And Signal Processing Temperature- and Supply Voltage-Independent Time References for Wireless Sensor Networks.

Stability measurements are only meaningful when the time between readings is specified. In other words, short- or long-term stability is the consistency of measurements taken at stated, equally-spaced intervals.

Revisiting applications such as microcontroller logic, serial communications, and some wireless communications up to the GPS time.

Intervals are rarely specified. While looking at Allan's deviation, it is a unit-less [%, ppm, ppb ...] quantity as long as the time interval is not specified.

On the right-hand side of the slide, typical counter results are given by a 24 MHz Crystal oscillator.

A universal counter (Keysight created 53230A) was used to data log frequency with a gate time of 1us

The first measurement of a frequency rising edge after rising edge is already integrated as available gate time of a standard counter and is rather a 1-μs period than 41.6ns, which is the expected value of the period of a 24 Mhz oscillator.

The short-term precision- so period-to-period measured error, in that case is 0.03 ppm while if a precise second is need, if it's about time-keeping, then a precision of 0.0003 ppm can be attained. This is extracted from the Allan deviation.

Wired and wireless communication comes with baud-rate providing an idea of the number of bits per second, from low data-rate to high-speed communication.

Evaluating frequency accuracy depends on the application, which is what the diagram on the left is showing.

The referenced values mostly demonstrate short-term precision, which is a must.

As the presentation is focused on time keeping, the most important metric is how precise the second is. In other words, what is the Allan variance after a second or what is the floor value and after how long is it reached?

Doing time-keeping for a duty-cycled radio would rather tighten that specification to within 1ms accuracy.

An example of the time interval is that between two data transfer events (BLE connection events) ranging from 7.5 ms to 4 secs, balancing throughput and power consumption.

The standard requires a maximum deviation of +- 500ppm.

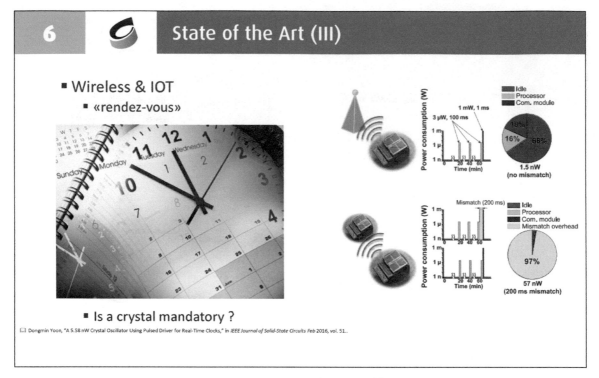

6 State of the Art (III)

- ■ Wireless & IOT
  - ■ «rendez-vous»
- ■ Is a crystal mandatory ?

Dongmin Yoon, "A 5.58 nW Crystal Oscillator Using Pulsed Driver for Real-Time Clocks," in *IEEE Journal of Solid-State Circuits Feb 2016*, vol. 51..

A full-duplex always on radio link for each and every sensor node won't make sense

The concept of a rendezvous, or meeting, is widely used in a group that exchanges information.

Translated in radio-communication technology, a standard device to gateway and internet such as IEEE 802.11 (Wifi) and IEEE 802.15 (Bluetooth) uses energy-efficient Time Division Multiple Access (TDMA).

Some negotiated parameters at the pairing are, for instance, the time between now and the next rendezvous.

This requires the chip to have a time base, on one hand. On the other hand, reserving a crystal oscillator footprint at a higher frequency for the PLL reference would generate a carrier frequency that is in the GHz range.

The study presented in the Journal of Solid State Circuits in February 2016 by Dr Yoon exemplifies that two independently-operating timing references will suffer from mismatch because of random behaviors such as jitter.

Timing reference, operating in sub-nW to realize a timer, reported Allan deviation of the leakage-based timer is three decades higher than a regular crystal oscillator.

This means two nodes are expected to have a random mismatch of 0.2 seconds in 3600 seconds, instability which corresponds to 56 ppm. For a point of comparison, a regular wristwatch is as accurate as 0.0025 ppm in an hour without using a pulse inhibition technique.

The consequence is that the RF communication module of one node will be activated earlier than the other and has to wait until the second node comes online before it can initiate communication.

During this mismatch time window, the RF module consumes substantial energy, which comes to dominate the total power usage.

To avoid this high energy loss during timer mismatch, each node must be equipped with a high accuracy timing source.

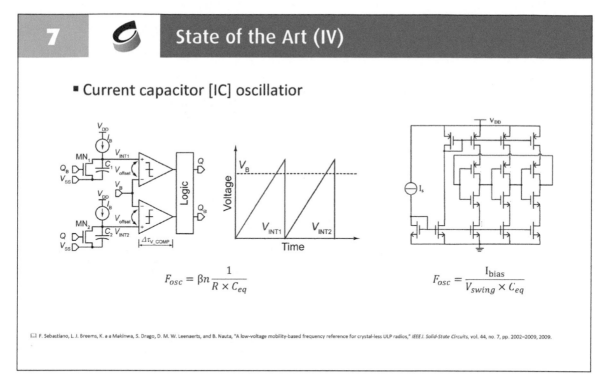

7 State of the Art (IV)

- ▪ **Current capacitor [IC] oscillatior**

$$F_{osc} = \beta n \frac{1}{R \times C_{eq}}$$

$$F_{osc} = \frac{I_{bias}}{V_{swing} \times C_{eq}}$$

▭ F. Sebastiano, L. J. Breems, K. a a Makinwa, S. Drago, D. M. W. Leenaerts, and B. Nauta, "A low-voltage mobility-based frequency reference for crystal-less ULP radios," *IEEE J. Solid-State Circuits*, vol. 44, no. 7, pp. 2002–2009, 2009.

A conventional voltage mode RC oscillator. Capacitors C1 and C2 accept bias currents IB and generate ramp voltages of V_{INT1} and V_{INT2}, alternately. By comparing the voltages with a reference voltage VB, the circuit generates clock pulse.

However, the clock frequency degrades due to the comparator's non-idealities , such as offset voltages and delays, as shown in this figure.

To improve the accuracy, several circuits employing a compensator have been reported.

They achieved high accuracy, which was 38-120ppm/°C.

However, they still dissipate high power, a few hundred nW.

Relaxation oscillators remain of high interest in research as their variability is related to capacitance which is small in the CMOS process, they have a resistor which is easy to trim and their mobility which is mostly temperature-sensitive.

Another interesting oscillator is the ring oscillator, the direct voltage control of which is very sensitive and direct and pulsed current tends to make power rails very noisy and are carried along the whole circuit. The current starved approach was first adopted in 1991 by Hewlett Packard as a patented technique US5072197 "Ring oscillator circuit having improved frequency stability with respect to temperature, supply voltage, and semiconductor process variations", where the switching delay was then related to the current, voltage and equivalent capacitance. Determining the equivalent capacitance and voltage swing given a current poses the biggest problem. Normally, the oscillators are slowed down by putting a load capacitor on each inverter output, and the linearity of the capacitor is a must. The bias current could be made proportional to a resistor so there is a first-order RC time constant relation.

These oscillators demonstrated low power consumption by scaling the voltage swing under the condition that the clock extraction or level shifter consumes a reasonable amount of current.

The following investigation has considered current controlled oscillators even though any node in the IC is capacitive and then a current encountering a capacitor generates a potential, so a voltage. Therefore, a current-controlled oscillator is a kind of misnomer.

8 Requirements

▪ Architecture requirements

Type	CMOS RC oscillator [1]	
Implement	Fully on-chip ! CMOS LSI chip PCB	
On-chip	Yes	
Power [nW]	4.7 nW	5 nW
Frequency [Hz]	3 kHz	8.192 kHz
Accuracy [ppm]	> ±1000 ppm	< ±1000 ppm
Temperature range [°C]	-25°C – 85°C	-25°C – 85°C
Supply range [V]	0.85 – 1.4	0.6 – 5
Supply sensitivity	0.48 %/V	< 0.5 %/V

[2] T. Jang, M. Choi, S. Jeong, S. Bang, D. Sylvester, and D. Blaauw, "A 4.7nW 13.8ppm/°C self-biased wakeup timer using a switched-resistor scheme," *2016 IEEE Int. Solid-State Circuits Conf.*, pp. 102–103, 2016

Although, a crystal oscillator is widely used because it shows high accuracy and power consumption is extremely low, it is adding a component to the bill of material, as a consequence adding additional cost.

IoT must be cheap in order to succeed, integrated circuit designers must succeed at building a crystal-less time reference.

It also increases the footprint of the whole system and, finally, it still has a decade higher consumption with respect to recent RC oscillators reported.

Some systems may require less accuracy over the temperature. There is, for instance, a digital inhibition scheme enabling compensation for a negative or a positive frequency offset.

These algorithms being developed for 32,768 Khz have taken advantage of the fact that the frequency is a power of 2 with a capability close to a multiple enabling the power of 2 computation.

But currently-reported architectures are limited to some consumer electronics and the temperature range (-25°C – 85°C) must be extended to an industrial application range (-40°C – 85°C).

RC oscillators for time keeping could be of interest in applications that run in a subthreshold microcontroller down 0.6V up to some applications that are battery-operated with a large spread. For instance, a system running with dry cells where the supply voltage is as low as 0.85V, while one using a lithium battery would require a capability of up to 5V . Also, some automotive applications would still be supplied under 5V supply bus.

Architectures reducing sensitivity to supply must be investigated.

9   Architecture consideration – Overview

VDD OSC

M1   Vrefp   M2

Ires   Iring

Vres   Vring

R   Vos

Rtrim   Cr   RING OSC

- **Under assumption of ideal components :**

- **Static current :** $I_{res} = \dfrac{V_{res}}{R}$

- **Dynamic current :** $I_{ring} = F \times C_{eq} \times V_{ring}$

- **Feedback loop :** $V_{ring} = V_{res} + V_{os}$

- **Feedback loop :** $I_{res} = I_{ring}$

$$F_0 = \frac{1}{R \times C_{eq}} \times \frac{V_{res}}{V_{ring}} \qquad F_0 = \frac{1}{R \times C_{eq}} \times \left(1 - \frac{V_{os}}{V_{ring}}\right)$$

With respect to state of the art technique, it is clearly a current controlled ring-oscillator even if it's a kind of misnomer as it is clearly represented a capacitance Cr that makes any I_{ring} converted into a voltage.

On one hand the ring oscillator average consumption is proportional to it's voltage, equivalent capacitance and Frequency.

On the other hand it exist a linear element that current is inversely proportional to the voltage over it's value, which is the resistor.

So it is proposed to put a current mirror(M1,M2), so that the DC current flowing into an on-chip resistor and the dynamic current supplied to a ring oscillator are made equal.

It won't be true as long as the drain voltage is subject to change and mismatch.

In principle, we could match the drains voltages equality by means of an OTA.

This proposed solution can as well suffer from the OTA input offset, which is represented by the voltage source Vos. The oscillation frequency is then derived and found that the oscillator is intrinsically dependent on a RC product, an offset voltage and the voltage itself. As a matter of fact, oscillation frequency rely on absolute values which could be trimmed in production, as the capacitance is defined by the inverter itself.

For a first order calculation the offset over the ring voltage can be neglected.

Implementation of the OTA can be discussed, as a source coupled comparator could also equate the two voltage. A self biased OTA was choosen for voltage compliance down 0.8V.

It's important to note that supply rejection ratio in current mirror could also be discussed as the most know and common results is about the output conductance in that specific case the power supply rejection ratio can be derived as :

$$PSRR_{V/V} = \frac{I_{ring}}{VDD\ OSC} = \frac{1}{r_{ds\,M2}} \times \left[1 - \frac{gm_{M1\|M2}}{gm_{M1\|M2} + \frac{1}{R} + \frac{1}{r_{ds\,M2}}}\right]$$

And as in first approximation trans-conductance is way larger than the admittance of the frequency setting resistor :

$$g_m \gg \tfrac{1}{R} + \tfrac{1}{r_{ds\,M2}}$$

It turns out to have an infinite supply rejection.

Which make this oscillator a great candidate to embed in complex System on chip.

The ring oscillator and resistor match their I-V characteristics in a one signle operating point.

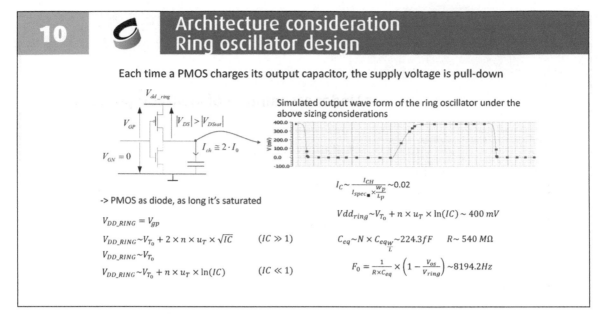

10 Architecture consideration
Ring oscillator design

Each time a PMOS charges its output capacitor, the supply voltage is pull-down

Simulated output wave form of the ring oscillator under the above sizing considerations

-> PMOS as diode, as long it's saturated

$$V_{DD\_RING} = V_{gp}$$

$$V_{DD\_RING} \sim V_{T_0} + 2 \times n \times u_T \times \sqrt{IC} \qquad (IC \gg 1)$$

$$V_{DD\_RING} \sim V_{T_0}$$

$$V_{DD\_RING} \sim V_{T_0} + n \times u_T \times \ln(IC) \qquad (IC \ll 1)$$

$$I_C \sim \frac{I_{CH}}{I_{spec_\blacksquare} \times \frac{W_p}{L_p}} \sim 0.02$$

$$Vdd_{ring} \sim V_{T_0} + n \times u_T \times \ln(IC) \sim 400 \, mV$$

$$C_{eq} \sim N \times C_{eqw} \sim 224.3fF \qquad R \sim 540 \, M\Omega$$

$$F_0 = \frac{1}{R \times C_{eq}} \times \left(1 - \frac{V_{os}}{V_{ring}}\right) \sim 8194.2 Hz$$

Each time a ring stage is in transition it's pulling down it's supply voltage into the load capacitor, thus behave as a diode as long at it stays in saturation. A model is proposed with the depandance of inversion level for V_{ring}, supply voltage of the ring oscillator as shown on slide 9.

With the knowledge of Equivalent capacitor and resistor it is then derived expected frequency.

11 **Temperature compensation**

The thermal behaviour of this circuit is expected to be fairly dominated by variations of the thermal coefficient of the passive elements R and C. MOS capacitances were -7300 ppm/°C which is too high to achieve stable oscillation frequency over temperature. For the resistor, it is a trade-off between high resistivity and thermal behaviour. A highly resistive poly resistor with a thermal coefficient of -1950 ppm/°C is also too high to achieve stable oscillation frequency for a time reference. To get a complete picture of the thermal dependency, the corresponding first-order relative temperature coefficient [ppm/°C] of the oscillation frequency is then given by a Tc equation.

The temperature coefficients R and C are expressed in ppm/°C, and temp coefficients of OTA offset voltage and ring oscillator voltage are expressed in µV/°C.

The above result shows that a first-order compensation of the temperature drift can be achieved under specific conditions.

Without any additional current branch or circuitry, it is possible to compensate the thermal behaviour.

In a perfectly-matched load, there remains two uncorrelated phenomena: random mismatch on threshold voltage and geometrical ratio. Therefore, the offset voltage was derived as Vos.

A differential pair implementing a structural offset was implemented in a binary weighted fashion (M8.i, M9.i) additional W, which is between 0 and 31 times more than the other side of the differential pair, corresponding to some frequency step in the order of 17Hz per step but flattening the temperature coefficient.

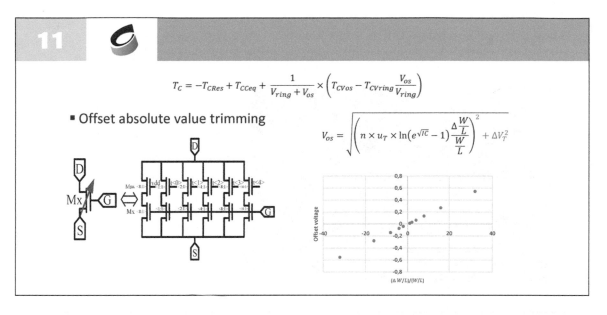

11

$$T_C = -T_{CRes} + T_{CCeq} + \frac{1}{V_{ring} + V_{os}} \times \left(T_{CVos} - T_{CVring} \frac{V_{os}}{V_{ring}} \right)$$

▪ Offset absolute value trimming

$$V_{os} = \sqrt{\left(n \times u_T \times \ln\left(e^{\sqrt{IC}} - 1 \right) \frac{\Delta \frac{W}{L}}{\frac{W}{L}} \right)^2 + \Delta V_T^2}$$

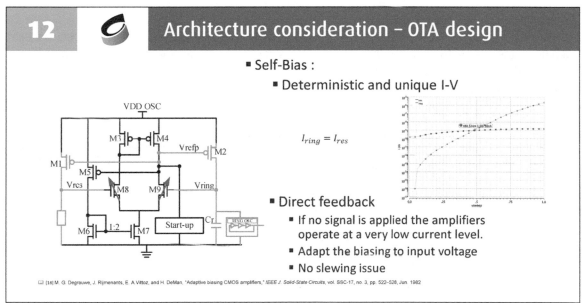

12 **Architecture consideration – OTA design**

▪ Self-Bias :

  ▪ Deterministic and unique I-V

VDD OSC

$$I_{ring} = I_{res}$$

▪ Direct feedback

  ▪ If no signal is applied the amplifiers operate at a very low current level.
  ▪ Adapt the biasing to input voltage
  ▪ No slewing issue

[16] M. G. Degrauwe, J. Rijmenants, E. A Vittoz, and H. DeMan, "Adaptive biasing CMOS amplifiers," *IEEE J. Solid-State Circuits*, vol. SSC-17, no. 3, pp. 522–528, Jun. 1982

Key feature for trimming implemented in the differential pair

An oscillator is most likely embedded in a complex system on chip where some other blocks tend to generate dynamic IR drop.

Accordingly, the sensitivity of the oscillator to a supply voltage variation should be minimized. Therefore, the selection of the OTA is then a very important point. An adaptive self-biased amplifier topology [16] has been proposed. An OTA has been designed (Fig. 4.) with the constraints of matching all the PMOS (M3, M4, M5) to the PMOS (M1, M2), so that the current biasing of the amplifier is a fraction of the load currents. Following the guidelines of [16], a ratio of 2 between M6 and M7 in order to maximize the gm/id and acheive a very high DC that almost zeros the input impedence (virtual ground), enabling $V_{Ring}=V_{Res}$. Any supply dependant offset voltage variation between the amplifier inputs is hence minimized. In order to reduce the start-up time, a start-up circuitry is introduced in order to trigger the positive feedback and suppressed it once the circuit is self-biased. Bias Ripple is mitigated by inserting capacitance at the gate of M8 and M9 (Cripfilt1, Cripfilt2).

13 Measurement results

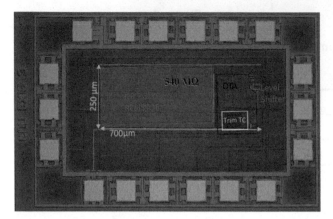

- 52 samples measured
- Automated Test Equipment
- Current consumption :
  - Average value (Pico-Amp)
  - Include core and level shifter

The die picture shows that despite choosing highly resistive polisilicon resistor, most of the area was taken up by the resistor (520µm x 250µm).

An important counter back of a high resistive poly resistor is it's thermal coefficient of -1950 ppm/°C which is an important penalty for temperature stability.

The test were carried in package on standard semiconductor automated test equipment's 52 Dils are tested, average value removing outliers is reported.

14 Measurement results – F=f(VDD)

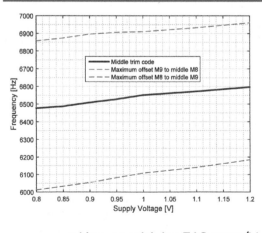

- Line sensitivity 710 ppm/V

The reported consumption was measure with a standard frequency counter (Keysight 53230A) with a gate time of 1us and is the average value of 52 samples.

This oscillator has a dependency of 710 ppm/V or 0.071% over the operating range 0.8V-1.2V Which is the best reported performance up to our knowledge.

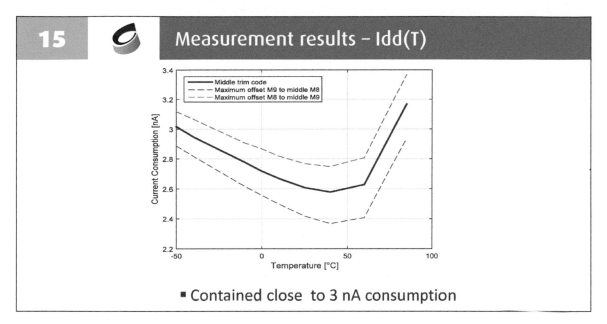

Measurement results – Idd(T)

- **Contained close to 3 nA consumption**

The reported consumption was measured with a calibrated Picoammeters (Keithley 6400) that provides average consumption and includes a core oscillator and level shifter.

The pad consumption is not included. This oscillator has shown a nA level operation in all its current branch over the 0.4V voltage range and the 135°C range.

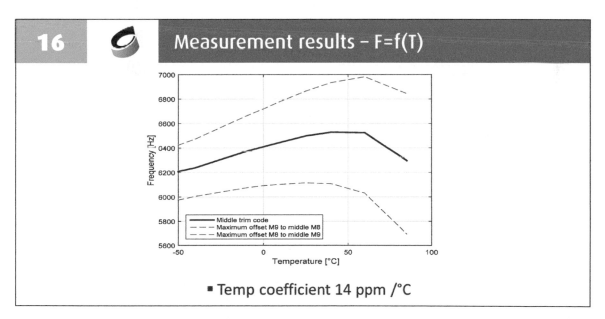

16

Measurement results – F=f(T)

- **Temp coefficient 14 ppm /°C**

The reported consumption was measures with a standard frequency counter (Keysight 53230A) with a gate time of 1us and is the average value of 52 samples.

In order to demonstrate the effective temperature calibration discussed, each trimming code was swept over the temperature and average frequency measured.

The different measurement results are presented following this convention.

The typical behaviour is presented without any imbalance as the bold line.

The maximum imbalance of the transistor M9 with respect to M8 is in red and the maximum imbalance of the transistor M8 with respect to M9 is in purple.

The calibrated imbalance has been selected as seven additional unity areas of the differential pair in order to get the best temperature coefficient.

The temperature coefficient is extracted as ±7ppm/°C in the temperature defined in usual human wearable electronics of [-40°C - 85°C].

In order to apply an eventual pulse inhibition, a more exact formulation of that coefficient would be a second-order polynomial function as the curve shape is parabola.

Before application, it is also important to characterize period-jitter

---

17 Measurement results – Jitter

- JESD65B suggest the following definition :
  - Measure each cycle period
  - Cumulate 10'000 consecutive samples
  - Plot an histogram

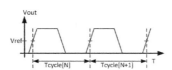

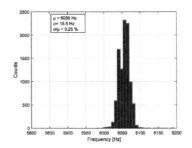

- **The period to period jitter is fairly gaussian.**
- **Integrating over a second the variability gets into the ppm range**

Jitter measurements were performed using a universal frequency counter, with 12 digits/s and a 20 ps resolution.

The JDEC standard JESD65B suggests the following definition: period jitter is defined as the maximum deviation of any clock period from its mean clock period.

It compares the length of each period to the average period of an ideal clock at a long-term average frequency of the signal.

Period jitter is typically specified over a set number of clock cycles. Jedec Specification JESD65B suggests measuring jitter over 10,000 cycles.

Every interval between rising edges was measured accordingly, and 10,000 measures were logged.

Then, it was decided to post-process by plotting a histogram, which is fairly Gaussian, so the first-order momentum could be considered as a standard deviation.

This mean inhibition technique can be applied under the conditions that the sampling phase is on average having the wished deviation .

A second reference would, for instance, not to deviate more than 12 PPM corresponding to one second being integrated per day.

18   Benchmark (I)

Parameter		This work	[10]	[5]	[6]
Process [nm]		180	130	180	60
Area [mm²]		0.175	-	0.24	0.048
Frequency [Hz]		6 000	100 000	11	32 728
T° range [°C]		-40 to 85	20 to 70	-10 to 90	-20 to 100
T° Coefficient [ppm/°C]		14	40	45	32.4
Line sensitivity [%/V]		0.071	-	1	3.5
Power [nW]	Un-comp	2,5	100	5.8	2800
	Comp	2,5	1000	5.8	2800
	Average	2,5	150	5.8	2800
Temperature compensation		1st order	yes	yes	yes
Principle		"CCO"	CCO	Relaxation	Relaxation
FOM : $\left[\frac{\mu A}{MHz}\right]$		0,42	1,5	527	85.5
FOM: $\left[\frac{\mu A}{MHz} \blacksquare\right] \times 10^6$		2.26	–	3903	1141

[10] A. Shrivastava and B. H. Calhoun, "A 150nW, 5ppm/??C, 100kHz on-chip clock source for ultra low power SoCs," in *Proceedings of the Custom Integrated Circuits Conference*, 2012, pp. 12–15.
[5] S. Jeong, I. Lee, D. Blaauw, and D. Sylvester, "Timer Using a Constant Charge Subtraction Scheme," pp. 3–6, 2014.
[6] K. J. Hsiao, "A 32.4 ppm/??C 3.2-1.6V self-chopped relaxation oscillator with adaptive supply generation," *IEEE Symp. VLSI Circuits, Dig. Tech. Pap.*, no. 1, pp. 14–15, 2012.

A lot of attempts have been reported to compare oscillators' efficiency such as the frequency over consumption figure of merit, which is similar to the power delay introduced for digital gates FOM1.

Some more recent concerns include area FOM3, defined as FoM1·Area/Lmin 2, where Lmin is the minimum gate length of the process technology.

And finally, the IoT thrust, to consider every SoC building bloc consumption as an energy per operation FOM2.

19   Benchmark (II)

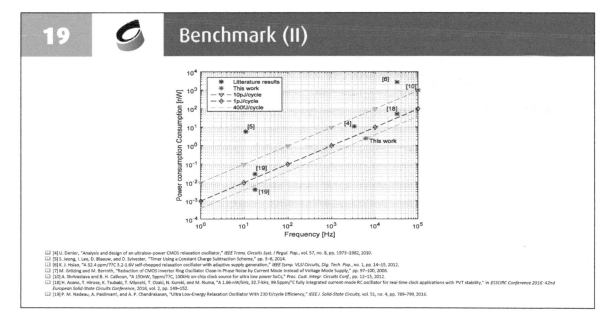

[4] U. Denier, "Analysis and design of an ultralow-power CMOS relaxation oscillator," *IEEE Trans. Circuits Syst. I Regul. Pap.*, vol. 57, no. 8, pp. 1973–1982, 2010.
[5] S. Jeong, I. Lee, D. Blaauw, and D. Sylvester, "Timer Using a Constant Charge Subtraction Scheme," pp. 3–6, 2014.
[6] K. J. Hsiao, "A 32.4 ppm/??C 3.2-1.6V self-chopped relaxation oscillator with adaptive supply generation," *IEEE Symp. VLSI Circuits, Dig. Tech. Pap.*, no. 1, pp. 14–15, 2012.
[7] M. Grözing and M. Berroth, "Reduction of CMOS Inverter Ring Oscillator Close-In Phase Noise by Current Mode Instead of Voltage Mode Supply," pp. 97–100, 2006.
[10] A. Shrivastava and B. H. Calhoun, "A 150nW, 5ppm/??C, 100kHz on-chip clock source for ultra low power SoCs," *Proc. Cust. Integr. Circuits Conf.*, pp. 12–15, 2012.
[18] H. Asano, T. Hirose, K. Tsubaki, T. Miyoshi, T. Ozaki, N. Kuroki, and M. Numa, "A 1.66-nW/kHz, 32.7-kHz, 99.5ppm/°C fully integrated current-mode RC oscillator for real-time clock applications with PVT stability," in *ESSCIRC Conference 2016: 42nd European Solid-State Circuits Conference*, 2016, vol. 2, pp. 149–152.
[19] P. M. Nadeau, A. Paidimarri, and A. P. Chandrakasan, "Ultra Low-Energy Relaxation Oscillator With 230 fJ/cycle Efficiency," *IEEE J. Solid-State Circuits*, vol. 51, no. 4, pp. 789–799, 2016.

The power delay introduced for the digital gates figure of merit introduced on the previous slide as FOM1 helps to benchmark our results against that of other publications in regards to time keeping.

It could be observed also then that the energy per operation figure of merit introduced as FOM2 is implicitly plotted.

It could be observed with respect to FOM2 that one paper reports a result without temperature consumption that breaks the 400fJ/cycle consumption, but without temperature compensation, which means it can no longer offer a frequency reference.

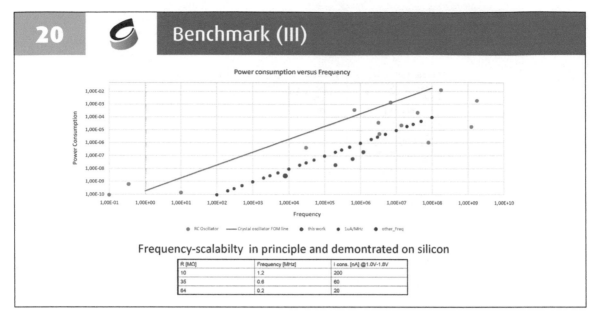

20 Benchmark (III)

Power consumption versus Frequency

Frequency-scalabilty in principle and demontrated on silicon

R [MΩ]	Frequency [MHz]	I cons. [nA] @1.0V-1.8V
10	1.2	200
35	0.6	60
64	0.2	20

As observed, there a quite a few time-keeping oscillators that are in a higher frequency range.

A few corrections were made to the OTA design and resistor scaling to generate higher frequency.

It could be observed that the figure of merit of 420pA/Khz could be improved down to 200pA/Khz instead, equivalently 200fJ/cycle.

This would constitute a further improvement.

21 Conclusion

- **LOWEST POWER PER HZ TOPOLOGY**

- **BEST THERMAL STABILITY**

- **Key features :**
    - Self biased ("PVT robustness")
    - Analog temperature compensation
    - Frequency scalability (From kHz to MHz range)

P. M. Nadeau, A. Paidimarri, and A. P. Chandrakasan, "Ultra Low-Energy Relaxation Oscillator With 230 fJ/cycle Efficiency," IEEE J. *Solid-State Circuits*, vol. 51, no. 4, pp. 789–799, 2016. [18Hz – 1kHz] (357fJ/Cycle – 166fJ/Cycle) [6k – 200k – 1.2M]

This work presents an oscillator that can be used in compact wireless sensors for time keeping mostly.

It describes a novel compensation strategy for thermal behaviour of an RC oscillator that could replace power-hungry solutions with a reasonable die-size.

A figure of merit expressed in µA/MHZ gives an indication of the spent power for a given activity as used with MCU.

I hope I convinced you that it's all about time and this architecture is the way to go.

CHAPTER 09

Low Power Wireless Embedded Systems

Marcel Meli

University of Applied Sciences
Winterthur, Switzerland.

Nodes for the Internet of Things are considered from a low power system point of view Some questions pertaining to energy and costs are addressed and examples of solutions proposed.

The examples shown are based on research work carried out by the research and development group of the author.

The contributions of various industrial partners and the CTI are acknowledged.

1   Outline

- **Generalities: Low-Power Wireless Embedded Systems**
- **Selecting components: Example of Bluetooth Smart TRX**
- **Cost effective harvesters**
- **Example of LEDs**
- **Powering LPWAN (here LoRa) systems with harvested energy**
- **Other examples**

This presentation has 4 parts:

- A quick look at the different parts found in a typical wireless embedded system that will help highlight important elements.
- Energy consumption: example of selection of components, especially in the case of Bluetooth

Smart.
- Energy generation: example of cost competitive harvesters such as LEDs/photodiodes
- A more demanding system. A LoRa node powered with harvested energy.
- The remaining slides show other examples of low power embedded systems.

2   A typical system.

- **Block diagram of a typical embedded wireless system**
    - The diagram shows the different elements found in a typical system
    - The system may be powered in different ways
        - Mains
        - batteries
        - Energy harvested from the surroundings
        - A combination

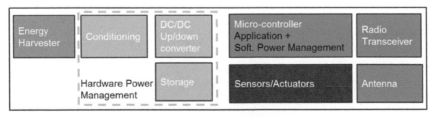

The system is powered by mains, batteries or energy harvested from the environment.

Different types of harvesters are possible. Combinations are also possible, depending on the application.

The power management strongly depends on the energy source / EH type.

The microcontroller is used for the application, but can also play an important role in energy management.

Sensors and actuators are chosen to fit the application.

The transceiver and the associated antenna should fit the frequency of communication.

3   Challenges

- **There are important challenges in designing systems that will be low-cost, maintenance free and energy autonomous for 10 to 20 years**
- **Harvester/storage elements are weak links in the chain**

From:
https://www.solarquotes.com.au/blog/solar-panel-cleaning/

In cases where power is partly or fully provided by harvested energy, harvesters and storage elements are weak links in the chain.

No work is possible when there is no energy.

- In some forms of harvesting, energy is not always available.
- Storage elements might be needed if the system should work when nothing Is harvested (e.g. solar cells at night).

More energy is required if there is more work to do.

- Harvester needs to deliver more energy.
- Generating more energy may require better/ larger/heavier harvesters.
- A better power management might also be required

Harvesters are often the most expensive part.

- They often require special power management.
- They can also be fragile (wear for harvester using mechanical parts, accumulation of dust on solar elements).
- Some harvesters are noisy, which is not acceptable in certain applications.
- Storage elements are also "fragile" and costly, especially if one needs to store a lot of energy.

4   Dealing with the challenges

Two approaches are helpful in dealing with the challenges:

Reduce the impact of harvesters and storage by reducing the energy requirements of the load.

- Use energy efficient system architectures (hardware and software).
- Better components

Reduce the cost of harvesting/power management.

- Reducing the energy requirements opens to door to other types of harvesters that may not generate much energy, but are low-cost and durable.
- We have investigated the suitability of LEDs/ photodiodes as harvesters for some use cases.
- We have also used low-cost noiseless piezo elements.

4

- **Reduce the impact of harvesters and storages**
  - Reduce the needs of the load
    - Hardware and software
  - Improve the efficiency of the energy conversion
  - Better (more efficient) power management
  - Go ASIC if necessary
- **Reduce the (cost) impact of harvesters**
  - We use LEDs in certain cases
    - Better power management needed
    - Low-cost elements
  - The piezo we use are easy to make and low-cost
    - Better energy transfer is needed

ENERGY REQUIREMENTS OF BLE TRX

5 | Energy required by components

- Comparing the Energy requirements of Bluetooth Smart solutions
  - Benchmark February 2016
  - Presented at Embedded World Conference 2016
  - More information available in paper/presentation
- Some devices require more energy
  - If possible, choose components that need less energy
- But first, we need to know how much energy the elements require

Since the reduction of the energy requirements is very important, one needs to use appropriate components.

The elements that are available on the market need to be evaluated. It is done here for Bluetooth Smart SoC.

- Part of the work done in February 2016 was presented at a the Embedded World Conference in Nuremberg.
- A new paper with the latest comparison is planned for 2018.

6		Energy requirements of Bluetooth Smart SoCs

- Motivation
  - Several Bluetooth smart solutions on the market
    - All claim to be low power, ultra low power, ... etc
  - Datasheet alone does not tell the whole story
    - Difficult to make comparisons
    - Difficult to derive needed information for energy estimation
  - Benchmark needed
    - Allows developers to quickly see important parameters
    - Allows fair comparisons between solutions (hard and soft)
    - Nothing existed for BLE, so we made one (2014, 2016, 2018)
  - Here is a summary of what we did

Bluetooth Smart solutions are complex.

- They cannot be clearly characterized just with the static measurements shown in a datasheet.
- A solution involves hardware and software.
- There are several activities within a SoC that determine the total energy consumption.

- The factor time is important in energy (i.e. how long does the clocks need to stabilize).

The motivation for this work is to gather the needed information in order to:

- Compare the different solutions that are on the market.
- Predict the energy consumption.

7		Procedure

- Procedure
  - Get as many devices as possible
    - Early sorting of devices (according to support / design date/ spe
    - Get dev kit from manufacturer when possible
    - BT stack from manufacturer (optimised if needed)
    - Measure devices under similar conditions
      - Dynamic power consumption in important modes
      - Same standard voltage, then VDDmin of each device as given by manufacturer
      - Averaging in some cases to reduce effects of DC/DC converters
  - Some manufacturers refused to give us their devices
  - More information can be found in paper

The support of manufacturers of Bluetooth Smart SoC solutions was requested. Some agreed.

Appropriates development kits and the needed firmware for the different solutions were used.

Basic modes were determined for the tests, dynamic power measurements made and the results entered in corresponding tables.

8   Energy requirements of BLE TRX

- Background Bluetooth Smart (work done for 4.1)
  - Short review on how Bluetooth Smart works
    - Advertisement (on up to 3 ADV channels)
      – Devices advertise their services
      – Scanners listen, possibly decide who to connect to
    - Connection
      – Following ADV, there can be request for connection
      – Parameters for connection are negotiated
      – After a connection, communication continues using data channels
    - Data can also be transferred with ADV (non connectable)
      – Example of beacons
    - Energy during coms will mostly go into: ADV (TX and RX), sleep, transitions between, clocking schemes, ... → to be measured

A small explanation of the way Bluetooth Smart works is necessary in order to understand the meaning of the tests.

More information can be found in the appropriate documentation (Bluetooth SIG standards).

Data can be exchanged in Advertisement mode or in Connection mode.

In ADV mode

Frames are regularly transmitted on 1 to 3 special channels, with information about the services offered by a device that is advertising.

In a special ADV mode (non connectable), the contains of the frame may be the data that the application broadcasts. This is the case for beacons.

Scanners regularly scans the ADV channels to get the advertisement information.

If a connection is needed with a specific advertiser, the connection parameters are first exchanged. Both devices then move to data channels to continue the communication.

In connection mode.

The 2 devices that are in connection mode meet at predefined times to exchange data.

Timing is very important here, since at the given rendez-vous time, one device will send and the other receive and vice versa. Otherwise both parties go in low-power mode to save energy.

9   Tools

- How we proceeded
  - Tools
    - Power Analyzer N6705B Keysight
      – 2 Quadrant PSU (N6781A)
      – Scope mode
      – Ranges: Automatic, 10 µA
      – Data Logger (Duration 60s)
    - Various sniffers, PCs, scope, RF-power detectors for verifications
  - Devices programmed for TX=0dBm (manufacturer's info)
  - Measurements
    - In Non-connectable and connectable ADV (frame with 31 bytes)
    - Use keep alive frames in Connected Mode (there is a TX and RX)

Tools.
- A Power Analyser from Keysight was used to record voltage and current and compute power/energy.
  - The instrument can automatically change ranges, which is important to cover the dynamic range of the SoC (mA to hundreds of nA).
- Various sniffers and oscilloscope were needed to verify that transmissions were successful.
- RF detectors were also used for verification.
- The appropriate development environments were required in order to change program parameters.

- Appropriate client devices were available for tests in connected mode.

Set-up.
- A maximal frame size with 31 bytes payload was chosen for the different ADV measurements (connectable and non-connectable).
- Data transmission was made with an output power of 0dBm, programming according to the manufacturer's datasheet (output power not measured).
- Keep-alive packets (empty) were used in connection mode.

---

10   Importance of chosen tests.

- **Cold start measurements**
  - Very important when working in beacon mode with intermittent Energy Harvesting sources
- **Advertising and connection events**
  - Energy for active event, sleep, cycle
  - Cycles of 100ms and 1000ms
- **Averaging (to average effects of DC/DC converters)**

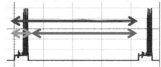

Cycle energy
Active energy / Sleep energy

Cold start measurements.

The device is powered up and energy measured between power-on and the end of the fist advertisement.

This is interesting for situations where an intermittent harvester is used (such as a push to harvest button).

Advertisement.

The device is programmed to advertise at a certain rate and the energy of a whole cycle is measured.

That energy is that of the active part (transceiver active) and the inactive part.

This is the main source of energy consumption for the beacon.

Before connection, a device must advertise.

Therefore, this measurement is also important for applications that require a connection mode.

Connection.

Measurements are made with the device receiving and then sending. The energy between the events is also measured.

If the device wakes up too early, the energy consumption of the receiver is increased, showing some timing issues in the solution.

Averaging.

This was used to average the effects of DC/DC converters on the measurements.

Several frames were captured over a given period and the energy divided by the number of cycles.

11 Tables

The devices we tested are listed in the table: Results for ADV_NONCONN_IND @ 3V

Devices	Measurement Voltage (V)	ADV event cycle (ms)	Cycle Energy (µJ)	Active part Energy (µJ)	Sleep part Avg. current (µA)	Sleep part Energy (µJ)	Remarks
					Parameters		
DA14581	3	106	31.3	30.9	1.3	0.38	External Flash, DCDC enabled, RCX20
RL78/G1D	3	100	32.8	32.4	1.4	0.39	DCDC enabled, internal oscillator
RL78/G1D	3	100	31.9	31.6	1.1	0.32	DCDC enabled, XT1 enabled
CC2650	3	100	44.2	44.1	0.1	0.03	DCDC enabled, LF RCOSC
SPBTLE-RF	3	104	41.8	40.9	2.5	0.77	DCDC enabled, LSOSC
BTLC1000	3	108	28.7	28.2	1.4	0.38	DCDC enabled, XOSC32K
NRF52	3	109	33.9	33.3	2.0	0.62	Eng. B revision, DCDC enabled, LFXO
CY8C4247LQI	3	109	94.1	93.7	1.8	0.51	PSoC, External Crystal Oscillator ECO
DA14581	3	1003	33.3	29.1	1.4	4.05	OTP memory, DCDC enabled, RCX20
RL78/G1D	3	1001	36.7	32.5	1.4	4.29	DCDC enabled, internal oscillator
RL78/G1D	3	1000	35.9	32.2	1.3	3.89	DCDC enabled, XT1 oscillator
CC2650	3	1001	46.9	45.5	0.5	1.4	DCDC enabled, LF RCOSC
SPBTLE-RF	3	1002	48.1	41.0	2.4	7.3	DCDC enabled, LSOSC
BTLC1000	3	1009	32.9	29.0	1.3	3.96	DCDC enabled, XOSC32K
NRF52	3	1004	41.4	34.1	2.4	7.32	Eng. B revision, DCDC enabled, LFXO
CY8C4247LQI	3	1006	98.4	94.5	1.3	4.00	PSoC, External Crystal Oscillator ECO

After capturing the power profiles, the energy was computed and entered in tables.

An example is shown here, for the ADV_NONCONN_IND @ 3V test.

The measurements were made for cycle intervals of 100ms and 1000ms.

The exact cycle duration as measured on the tool is also written.

12 Example of power profile

- **Measured parameter:**
  - **Energy from cold start-up until first non conn ADV**
    - Significant for beacon with EH. Here, case Dialog DA14581 (@3V)
    - Start-up time: 31ms, Energy required: 113.1µJ (incl. ADV)
    - Blue → voltage, brown → current, green → power

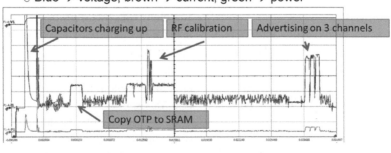

The different phases between power-up and the end of the first ADV can be seen.

This enables the manufacture or developer to see where there is a potential for energy reduction.

I2 is the current, V1 the voltage, P1 the power.

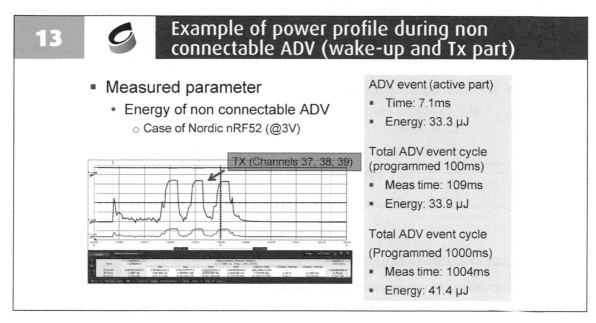

13 **Example of power profile during non connectable ADV (wake-up and Tx part)**

- **Measured parameter**
  - Energy of non connectable ADV
    - Case of Nordic nRF52 (@3V)

TX (Channels 37, 38, 39)

ADV event (active part)
- Time: 7.1ms
- Energy: 33.3 µJ

Total ADV event cycle (programmed 100ms)
- Meas time: 109ms
- Energy: 33.9 µJ

Total ADV event cycle (Programmed 1000ms)
- Meas time: 1004ms
- Energy: 41.4 µJ

Energy from the time the device wakes up, prepares to transmit, transmits on 3 channel, goes in low-power mode.

The current curves give an idea of the "internal activities" of the system, delivering information for potential improvements of HW and FW.

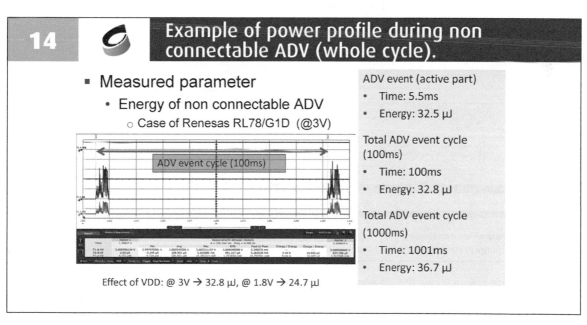

14 **Example of power profile during non connectable ADV (whole cycle).**

- **Measured parameter**
  - Energy of non connectable ADV
    - Case of Renesas RL78/G1D (@3V)

ADV event cycle (100ms)

ADV event (active part)
- Time: 5.5ms
- Energy: 32.5 µJ

Total ADV event cycle (100ms)
- Time: 100ms
- Energy: 32.8 µJ

Total ADV event cycle (1000ms)
- Time: 1001ms
- Energy: 36.7 µJ

Effect of VDD: @ 3V → 32.8 µJ, @ 1.8V → 24.7 µJ

Energy for a whole cycle is shown (active and inactive parts).

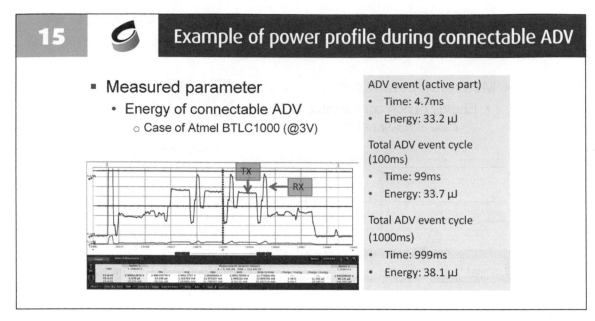

For each channel, the device transmits, then goes in receive mode for a short time in order to receive an eventual request from a scanner that wants to connect or wants more information.

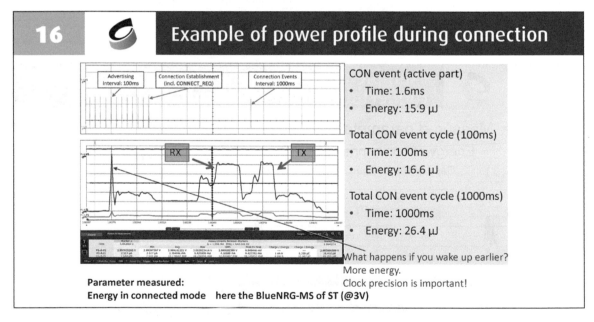

The first figure shows how the devices get into connection mode.

- First there ADV are transmitted, then connection parameters are exchanged. Finally, the devices go in connection mode.

The second picture highlights the connection part where the device wakes up for a rendez-vous, turns on its receiver to get a packet from the master and then transmits information to the master. This exchange may continue for a given number of packets. After that, the devices may go in low-power mode until the next appointment.

Energy while receiving and while transmitting normally dominates. if the connection interval is long, the low-power mode energy becomes very important.

17   Conclusions

- Conclusions
  - Energy consideration very important for battery operation and energy harvesting
  - A good analysis of the energy profile is important for manufacturers of ICs and application engineers
  - Optimal device depends on specific use case
  - Latest generation of Bluetooth smart chips with strong improvements compared to the last one
  - Full comparison listed in paper (February 2016)
    - New comparison paper foreseen for first half of 2018

Energy consumption of the load is important in the design of low-power systems.

Information about that consumption is needed in order to chose the appropriate components.

This work evaluates the energy requirements of Bluetooth Smart SoC and classifies that information.

That helps the process of choosing components and estimating energy requirements for an application.

USING LEDS TO HARVEST ENERGY

18   Introduction

- Wireless sensing using LEDs as very low-cost Energy Harvesters
  - How do you lower the cost of EH powered wireless nodes?

Energy Harvesting is interesting but still expensive, especially compared to systems that use batteries as power source.

This cost is an obstacle in certain important markets.

In this part, we consider the use of low-cost harvesters in an attempt to reduce cost.

The work was carried out in 2014/2015 and presented at the Embedded World conference in 2015, in Nuremberg.

19 — Energy Harvesting is expensive. Can LEDs be used as low-cost harvesters?

- **Motivation**
  - EH is often expensive compared to batteries
  - Some reasons
    - ○ Extra electronics for appropriate PM to deal with the type of EH
    - ○ Requires more volume to get the quantity of energy needed
    - ○ Needs storage to make up for "bad days"
    - ○ Harvesters are still expensive. Low volume → costs remain high
  - What can be done?
    - ○ Simplify the elements in order to use standard circuits
    - ○ Use "low-cost harvesters"

Although the idea of harvesting energy is attractive, it has not yet really taken off in terms of volume (except for some special cases such as calculators powered by small solar cells). One reason is that energy harvesting is still expensive. Why?

- Harvesters are not (yet) produced in large quantities (as batteries are) and remain expensive.
- Special electronics is sometimes needed, depending on the harvester type that is used. This means additional components.
- In some cases, energy must be stored in order to use it during the «bad days», when there is little energy around.
  - In such cases, storage elements and the associated power management are needed, adding to the cost and complexity of the system.

The main motivation for this work was to find out if LEDS (and photo-diodes) could be used as low-cost harvesters, especially for wireless embedded systems.

An important objective was also to determine the conditions under which such a system could work.

20 Why LEDs?

- Are produced in large quantities
- Have been used as sensors. Convert light into electrical energy under certain conditions
- Small size
- Mounting on PCBs is "mastered"
- Low cost → some will costs 1-2 cents in large quantities
- But can they work as EH in some of today's systems?
  - How much energy? How easy to harvest? (how complex are the special circuits?)
  - In which applications could they make sense?

LEDs are normally used to produce light. However, they can and have been used as sensors.

They are produced in very large quantities today, which has contributed in reducing their costs.

LEDs are used in several electronic applications and therefore well integrated in manufacturing processes.

They occupy a relatively small volume, compared to standard coin batteries.

Photo-diodes are made for sensing and will normally harvest more energy that LEDs.

We therefore concentrate on LEDs here.

21 A realistic application

- **Defining a realistic application**
  - Define an application to keep the work application centered
  - Important points of the application:
    - Measure a slow varying parameter. Temperature in garden.
    - Use LEDs in the appropriate environment as power source. Normally, there is enough light in the garden during the day
    - Use a popular wireless system to communicate information. We use Bluetooth Smart in ADV mode

In order to evaluate the use of LEDs as harvesters, we defined a realistic application.

- Harvest energy with LEDs in order to measure parameters such as temperature in a garden.
- The chosen parameter varies very slowly,

therefore one can afford missing some frames or measuring at longer of minutes.

- The measurement results are sent to a smartphone with Bluetooth Smart interface and the appropriate app for display.

22   Set-up

- **Set-up for measurements (enough energy?)**
  - ○ 3-4 LEDs used (yellow, red) with different angles → keep costs low
  - ○ Scope and high impedance probes for measurements
  - ○ Luxmeter, Power Analyser, light sources
  - • ES (Low power ASICs, sensing, capacitor as storage)
    - ○ The load can run on < 50µJ (3 frames BLE), < 20µJ (proprietary)

A set-up was prepared to help estimate the amount of energy that could be harvested under different light conditions.

This was done to have an idea of the amount of energy scavenged. The illumination measurements are not precise.

The luxmeter gave us an estimation of the illumination.

A board with LEDs and a load allows us to test different combinations of LEDs (series or parallel).

The oscilloscope was used to measure the amount of energy in the storage capacitor.

The load uses some special components (ASICs) in order to reduce the energy consumption.

23   Load requirements.

- **How much energy do we need?**
  - ○ Measurement on one of our optimised system
  - ○ Startup , measure, send Ble frame (Preamble .... CRC →29 bytes)
  - ○ Our system: needs at least 12.3µJ (only 1 Frame!)

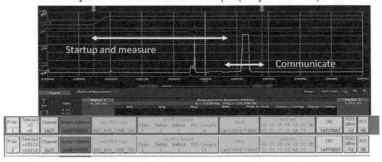

23

Knowing the load requirement is important for the design of the system.

We first optimised one of our wireless nodes and then measured the energy requirements.

The picture shows the energy profile of a systems transmitting Bluetooth Smart frames.

The frame structure shows all the bytes that are sent. More data than necessary is sent. However, the overhead is needed for compatibility.

If only one ADV frame is sent at a stable low voltage, about 12.3 mJ is required from startup.

Sending more frames requires more energy, but increases the probability that the frame will be seen.

If a dedicated receiver is used, one could further optimised the frames and reduce energy consumption.

24 **Results (BLE ADV)**

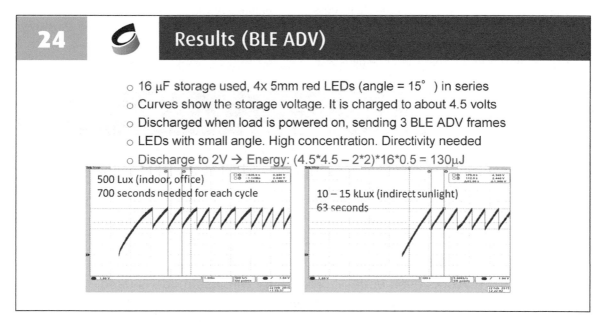

o 16 µF storage used, 4x 5mm red LEDs (angle = 15°) in series
o Curves show the storage voltage. It is charged to about 4.5 volts
o Discharged when load is powered on, sending 3 BLE ADV frames
o LEDs with small angle. High concentration. Directivity needed
o Discharge to 2V → Energy: (4.5*4.5 – 2*2)*16*0.5 = 130µJ

500 Lux (indoor, office)
700 seconds needed for each cycle

10 – 15 kLux (indirect sunlight)
63 seconds

Results when using a configuration with 4 red LEDs in series, and a 16mF capacitor to accumulate the energy.
- The storage charges to 4.5 volts.
- The embedded system is connected.
- The microcontroller starts up, measures, configures the radio and loads the data to send.
- Data is sent in Bluetooth Smart ADV frames

- The storage capacitor is emptied down to about 2.2 volts
- The energy accumulation process restarts.

Between 4.5 volts and 2.2 volts, 130mJ energy is released to power the embedded system.

The energy requirements of the load are clearly important in the performance of the system.

25 Results (proprietary wireless -> less energy)

- o 2 µF storage used (less energy stored)
- o 4x 5mm red LEDs in series. Curves show storage voltage
- o Storage cap is charged to about 4.5 volts. Discharged when the load is powered on, sending 1 proprietary wireless frame
- o Discharge to 2V →16.25µJ between 4.5V and 2V

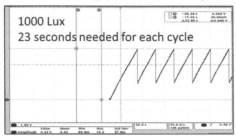

1000 Lux
23 seconds needed for each cycle

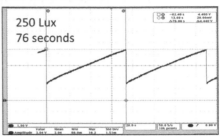

250 Lux
76 seconds

The curves show the voltage at the storage capacitor, in the case of a smaller capacitor (2mF).
- The system works under indoors conditions (less than 1000 Lux).
- Since the storage is smaller, less energy is needed to get to 4.5 volts. Less energy is also available for the load.
- The energy is however sufficient to transmit proprietary wireless frames.
- It should be remembered that the LEDs used here integrate a lens.

26 Conclusion

- ▪ Conclusion: Successful use of LEDs as EH to power simple wireless sensor.
    - 4 red LEDs are good enough (less than 10 cents)
    - Works using Bluetooth Smart ADV frames
    - Works well outside. Also works indoors
    - Works even better if proprietary wireless mode is used
        - o Most energy goes in the communication
    - Update rate of less than a minute possible
    - Potential cost is low (comparable to batteries)

We successfully showed that it is possible to use low-cost components such as LEDs as energy harvesters. This helps reduce the cost of energy harvesting. The 4 red LEDs cost less than 10 cents in total, in large quantities.

By using a low-power electronic switch and a capacitor, energy can be accumulated and delivered once there is enough for the task at hand.

The systems works well for the defined use case.

POWERING LORA WITH EH

---

27   **Why LP-WAN?**

- Powering LPWAN with harvested energy
  - Work done in 2016/2017
  - Presented at Wireless Congress, Munich
  - Papers/slides available
- Why LP-WAN?

Solar cell
20mm x 6mm

Designed by ZHAW-InES   Source: ZHAW

P-WAN (Low Power Wide Area Network) have become important elements in IoT.

- The coverage of Wide Area systems is much more than just a building (ranges of several kilometers are possible).
- Therefore, they facilitate the network architecture by reducing the number of gateways (compared to mesh systems with PAN radios)
- They allow energy efficient transfer of small amount of information over long distances (compared to cellular networks)
- Node cost, size are acceptable.
- Market intelligence forecasts hundreds of millions of LP-WAN nodes for IoT.

There are many LP-WAN systems. LoRaWAN has been deployed by Swisscom in CH.

There is however a fundamental question. How will these millions of nodes be powered?

---

28   **Motivation**

- How are LP-WAN nodes powered at the moment?
- There is a case for powering those nodes with harvested energy
  - But is that possible?
  - What are the issues and the limitations?
  - How can they be overcome?
  - This work looks at such issues

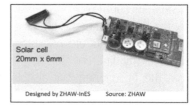

Solar cell
20mm x 6mm

Designed by ZHAW-InES   Source: ZHAW

ow are the nodes powered at the moment?

- Mains. But that negatively impacts mobility. Not possible is most cases.
- Batteries. Most examples / documents consider powering LoRaWAN nodes with batteries.
  - Mostly for Class A. Not good enough for other classes (require more energy).
  - Tens years or more using batteries? That is often said.
    - ~ It depends on use cases, types of batteries, range ...

- ~ The best range, SF12 mode, requires several hundred millijoules (see measurements later).
- Use of EH is an alternative. As main power source or support to prolong battery life.
  - But size and costs while using EH need to be kept low.
  - Careful design is required. Including for Power Management.

It is first important to understand where the energy goes.

A short background to LoRa (I)

- **Understand the requirements**
  - **Background to LoRa (868 MHz (EU), CSS modulation)**
    - UL: Variable payload size. Data rate depends on SF
    - Several spreading factors → different ranges
      - SF12 slowest + best range

Transmit	Receive	Receive
Uplink	window1	window2
		DL

delay1

delay2

After transmitting, the object might receive a DL after delay1 or after delay2.

  - Bidirectional mode possible
    - (variable payload size)
  - 2 reception windows with delays
    - delay1 is 1 sec, delay2 is 1 sec
  - 3 classes of devices. Class A is lowest power and interesting for us

---

oRa radios are presently used in one of the popular LP-WAN systems (LoRaWAN). Proprietary networks with LoRa are also possible,

Those networks have been (and are being) deployed in several countries.

In Europe, the 868 MHz band is used.

Several spreading factors are possible. They can be used in parallel (by different radios).

SF12 gives the best range, but also leads to a lower data rate. That results in long frames and thus more energy.

Bidirectional communication is possible, but limited. After the Up Link transmission, a device can receive a frame from the gate way in one of 2 windows. Of the 3 available device types supported by LoRaWAN, class A is the most popular and also the one best suited for low power applications.

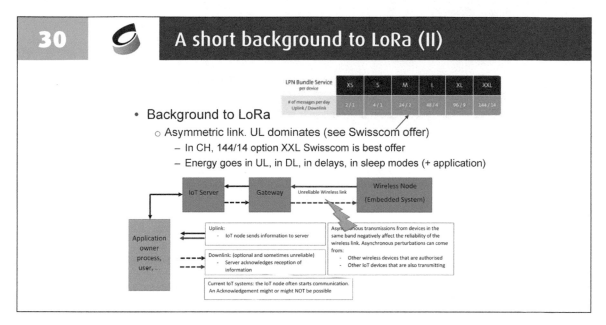

The use of the 968 MHz band imposes some limitations. Devices must respect a certain duty cycle while transmitting.

Although bidirectional communication is possible, the link is asymmetric.

In Switzerland (Swisscom network) for instance, a maximum of 144 frames per day can be sent by a node to the gateway. 14 downlinks at the most are allowed.

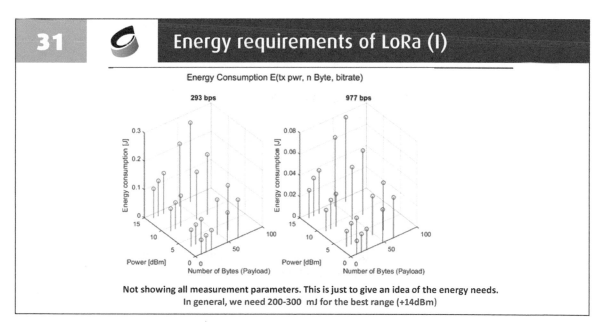

In order to estimate the energy needs, several measurements were carried out.

The results show how the energy requirements vary with the payload, the spreading factor and the radio output power.

It can be seen that for the best range (SF12, +14 dBm) one needs 100-300 millijoules.

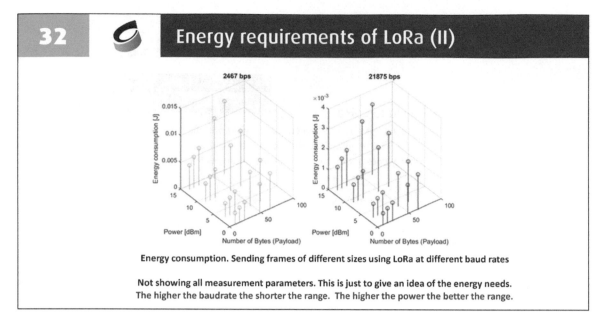

| 32 | | Energy requirements of LoRa (II) |

Energy consumption. Sending frames of different sizes using LoRa at different baud rates

Not showing all measurement parameters. This is just to give an idea of the energy needs.
The higher the baudrate the shorter the range. The higher the power the better the range.

The results show that the energy requirements decrease as the data rate of the communication increases.

But it should be kept in mind that this also leads to reduced ranges.

| 33 | | Energy requirements of LoRa radio |

- **Understand the requirements**
  - **Consequences for power supply (vs. other systems)**
    - o BLE requires 10s-100s of microjoules for tens of bytes
    - o GSM → requires Joules
    - o LPWAN → millijoule to hundreds of millijoules
      - – However, the range is clearly better than that of WPAN
    - o Approximate comparisons LPWAN vs. others:
      - – Vs. WPAN: 100 to 1000 times more energy required
      - – Vs. GSM: 10 to 100 times less energy
  - **Vs. WPAN: peaks of current important (especially at +14dBm). Frames are longer (especially for SF12)**
    - o Harvester size and PM choices should take this into account

The system needs to be designed to sustain the longest possible range, that is SF12 at +14dBm.
This means that the power management/storage element should be able to deliver the energy required at the correct voltage for hundreds of milliseconds or even seconds, depending on the frame size.
This is a major difference with PAN system that require 100 to 1000 times less energy.

34 — The basic concept

- Basic concept
  - PM + support elements to improve the efficiency
  - DSSC from GCell

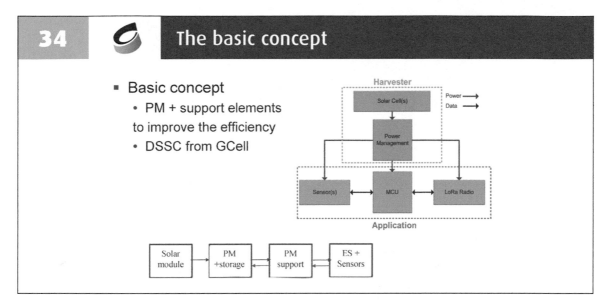

The basic concept includes typical elements found in a low-power wireless embedded system.

The power management is necessary to allow the accumulation of sufficient energy before the transceiver is started.

Energy is accumulated in a super capacitor and constantly monitored for the value necessary to guarantee the operation of the embedded system.

In case of a single solar cell, a booster (integrated in the PM) is needed to step the voltage up.

Since a LoRa frames can last for several hundred milliseconds, the power management may need extra elements to improve its efficiency.

These elements are represented in the block PM support.

35 — The final system

- System can work indoors and at window
  - PM + support elements to improve the efficiency
  - DSSC from GCell

Picture of the embedded system (solar cell, microcontroller and radio).
Solar cell: 167mm x 53mm x 0.5mm (active solar cell area is 140mm x 24 mm)
Sensors from Sensirion, Bosch, Analog Devices.
PM from EM , radio from Semtech, Micro from ST. e-peas PM also used successfully

The final system can measure diverse parameters thanks to on-board sensors and transmit the results using LoRa.

A DSSC (dye-sensitized solar cell) was used in 2 settings:

- At the window of one of our offices (mostly illuminated by outdoors light).
- Far from the window (office illumination conditions).

36 — Power management elements

EM8500
Cold start from 300mv, hot start from 100mv
Serial link for configurations
2 storage elements (STS and LTS)
Output line of 20mA @ 1.8V.
Several outputs could be combined to increase current (Max 50mA).

e-peas AEM10940
Current output as high as 80mA
Cold start from 380mv
Hot start from 100mv
11uW typically needed for cold start
Certain configurations are defined.
Other setting possible with external resistors

Two power management ICs were used at different stages of this work.
- The PM from EM Microelectronic (EM8500)

- The Power management of epeas (AEM10940)
 We added extra electronic to these elements in order to improve the energy performance.

37 — Results

- In SF10; Better than 144 UL per day (must be limited)
- In SF12; Tens of frames on a cloudy day (see below)

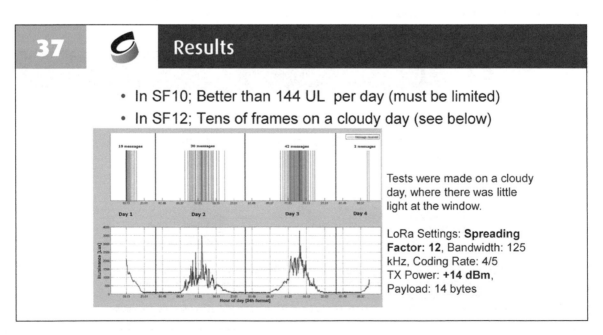

Tests were made on a cloudy day, where there was little light at the window.

LoRa Settings: **Spreading Factor: 12**, Bandwidth: 125 kHz, Coding Rate: 4/5 TX Power: **+14 dBm**, Payload: 14 bytes

The results for the case the solar cell was used at the window are shown.

A measurement giving an idea of the illumination is also shown.

It is possible to accumulate enough energy to reach the maximum of 144 frames per day.

Even on a cloudy day, at the window, the 144 frames are possible in SF10 mode.

Note that these are only UL frames. The system was not programmed to receive DL frames.

Refer to the published paper for other tests.

38 Conclusions and outlook

- Reduce the size of the solar cell to reduce costs and volume
  - Test for indoors and outdoors.
  - Store extra energy can be stored for periods with less illumination
- Use other harvesters
  - TEG, vibrations, ... etc
- Try other wireless systems

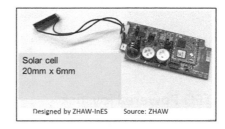

Solar cell
20mm x 6mm

Designed by ZHAW-InES   Source: ZHAW

The work yielded valuable information on aspects that could be optimised.

Having demonstrated that it is possible to power a LoRa node on an indoor solar cell, we will now turn our attention to reducing the size and cost of the system. Our future work will focus on using low-cost elements of smaller size and improving the power management.

OTHER LOW-POWER SYSTEMS

39 TEG used for EH

Source: ZHAW-InES

Receiver connected
to PC with game SW

Battery-free Wireless system
TEG with PM

TEG-Powered joystick

- Temperature differences between body and environment
  - Provides the needed electrical energy
  - The movements of the stick are measured and wirelessly sent to the PC
  - Update rate important → fast measurement/transfer of data for game

Battery-less Joystick

This demonstrator uses the energy harvested from the body heat (temperature differences between body and environment).

A TEG converts that energy into electrical power. It is then used to power a hand-help system that tracks the movements of the hand.

The sensor data is wirelessly sent to the PC for the control of the game.

40 Piezo used for EH

ALGRA GROUP

ALGRA ■ MICRODUL
Customized Swiss Microelectronics

Source for picture: Algra

Dynapic Wireless Presenter

- Innovative application of the Dynapic Wireless Piezo
- Use the piezo to provide energy to control a presentation (Power Point, etc.)
- Two piezo buttons for navigating through the presentation
- >10 Mio press/release cycles
- Cooperation
  - ZHAW-InES, Algra, Microdul

Dynapic Wireless Battery-less Presenter

The Dynapic Wireless is a piezo element made by the Swiss firm Algra.

It is flat, low-cost and can handle more that 10 million cycles. It only produces tens of microjoules when pressed and released.

The energy generated is used to power a microcontroller and a radio that will transmit a coded signal in 2.4 GHz band.

In this application, it is used to remote control a presentation (demonstrated during this conference).

- Pressing one key will move to the next slide.
- Pressing the other key will go back to the previous slide.

Work related to the Dynapic Wireless is a cooperation between Algra, Microdul and ZHAW, supported by the CTI grants: 18376.1 PFES-ES, 12705.1;5 PFES-ES.

---

41 — Piezo, RF harvesting

- **Press to measure and display**
  - Wireless communication not always needed
- **RF harvesting (868 MHz) to power BLE sensors**
- **RFID emulation with microcontroller (RFID sensors)**

Harvests from RF, communicates using BLE

Harvests from RF, communicates using RFID protocol

Source for all pictures: ZHAW-InES

Press and measure.

Measurements are started when the piezo is pressed. The energy from the Dynapic piezo (Algra) is enough to:

- Make a measurement, erase the e-ink, display the new value on the e-ink display.
- The information remains readable, even when there is no more energy in the system.
- The next press action will start a new cycle.

RF harvesting.

A dedicated RF source is used to power the embedded system.

The microcontroller and a Bluetooth Smart radio are used to make a measurement, process the data and sent it in ADV frames.

RFID emulation. Energy comes from the field of the RFID reader.

A low power microcontroller is used to emulate the behaviour of a UHF RFID Tag.

The system can sense some parameters and transmit them as passive RFID devices.

8-bit or 32-bit microcontroller were successfully used.

For copyright reasons, our speaker from Microsoft cannot share the content of the presentation. A very brief summary slide of that talk is proposed instead as a part of chapter 9.

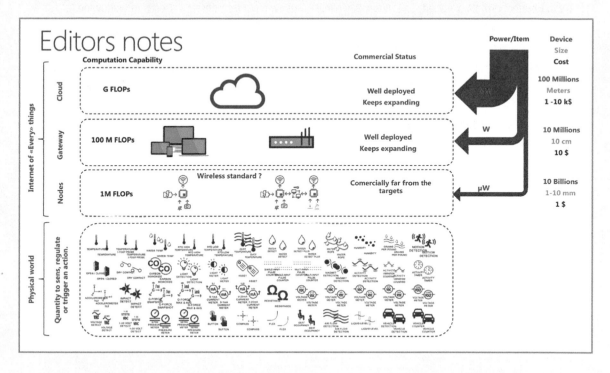

At that point of the workshop, the picture of Internet of things and everything is getting clearer. The combined sensor, processing and wireless communication had shown promising results.

The above figure attempt to summarize that there is such a large variety of quantity that could be sensed and regulated or trigger an actuation.

The Internet of everything or internet of things on itself could be considered as a large network of connected things, sensing or actuating their environments.

To do so, it's usually three to four layers are considered The Edge/Node, The gateways and the data-center/ cloud. In the middle of gateway and Cloud it is sometime considered an IT-Edge containing a processing and a analytics as some nodes doesn't feature any.

Order of magnitude are also provided regarding power per item, the number of devices, their size cost and computational capability.

Data-centres have been designed to allow operational and capacity changes and expansions and in conjunction with the fact that many of them run redundant power and cooling systems to provide better reliability, the energy consumption reported in 2012 world wide is about 269 Tera Watts. This was an important motivation to invite one of the key players active in the IoT cloud Buissness discussing obviously on their present Carbon footprint [2]. Although there are some regulations [3-4], some action could be mentioned [5-6]. Service provider have understood the need to harvest renewable energy and are committed to it.

From The IT perspective, there is no technological concept breakthrough. The Internet of Things isn't a technology revolution. IoT is a business revolution, enabled by technology. IoT is already delivering tangible results, several solution are exposed in [7].

References :

[1] Van Heddeghem, W.; Lambert, S.; Lannoo, B.; Colle, D.; Pickavet, M.; Demeester, P. Trends in worldwide ICT electricity consumption from 2007 to 2012. Comput. Commun. 2014, 50, 67–76

[2] https://www.microsoft.com/en-us/environment/carbon/default.aspx

[3] European Commission. EU Code of Conduct on Data Centre Energy Efficiency. Introductory Guide for All Applicants. Version 3.1.2. Available online: https://e3p.jrc.ec.europa.eu/publications/ict-code-conductintroductory-guide-all-applicants-v312

[4] International Organization for Standardization. ISO/IEC 30134-2:2016 Information Technology—Data Centres—Key Performance Indicators—Part 2: Power Usage Effectiveness (PUE); International Organization for Standardization: Geneva, Switzerland, 2016.

[5] https://blogs.microsoft.com/green/2016/10/27/energy-efficiency-and-designing-the-datacenters-of-the-future/

[6] http://download.microsoft.com/download/A/F/F/AFFEB671-FA27-45CF-9373-0655247751CF/Cloud Computing and Sustainability - Whitepaper - Nov 2010.pdf

[7] https://iotweek.blob.core.windows.net/slides2017/GIoTS/GIoTS%20Large%20Industries%20Views%20on%20IoT%20Data%20and%20Security/M.%20Janik%20Microsoft%20.pdf

Microsoft's Approach to IoT
Real-life examples of IoT as a platform service

Sherryl Manalo
Cloud Solution Architect
Microsoft Schweiz GmbH

The talk entitled "Microsoft's Approach to IoT Real-life examples of IoT as a platform service" is aimed at providing an interface between the circuits and things presented along the workshop. What does it take to connect the world's devices to the cloud, what are the common patterns across industries, how does the collected data provide real value? Processing the world's data requires a large infrastructure and consumes a considerable amount of energy.

All this infrastructure will consume power, and using IoT might be challenged with the question of it's global footprint. However, intelligent solutions in the cloud can help reduce energy use by up to 90% per user. Cloud infrastructure can play an important role in deploying clean and renewable energy. This helps maximize the benefits associated with data center transformation. For Microsoft, we are moving beyond datacenters that are already 100 percent carbon neutral to also having those datacenters rely on a larger percentage of wind, solar and hydropower electricity over time. Like many countries, we understand the task is not easy but we are confident that with smart investments, technologies and partners we can achieve this goal.

Microsoft has a common infrastructure for our cloud and online services. The common infrastructure is used by all of Microsoft's online and cloud services which span the Infrastructure, Platform and Software as a Services service models. Windows Azure and other Azure services deliver Infrastructure and Platform as a service both internally to other Microsoft services, as well as externally to Microsoft's customers and partners.

The **Internet of Things** isn't a technology revolution. IoT is a business revolution, enabled by technology.

IoT is already delivering tangible results, here are some examples:

Manufacturing

Moving from action to insights

Reduced development time and cost by 80% by gathering and analyzing data more efficiently and increasing automation across the company.

Improving equipment to weather the storm

Gained visibility into equipment usage and a path to predictive maintenance using machine learning.

Filtering the signal from the noise

Used analytics to discover actionable insights around fuel usage, predictive maintenance and stop unscheduled delays.

Smart Infrastructure

Transforming the urban landscape

Gathered data from sensors and systems to create valuable business intelligence and shift from reactive to proactive maintenance.

Empowering global operations

Collected data from more than 40,000 sensors around the world to create a best-in-class forecast that helps customers make critical operational decisions.

CHAPTER 10

Vertical Co-design and Integration in Energy Harvesting: from Device, Circuit and System Levels to IoT Applications

Eduard Alarcón

UPC BarcelonaTech
Barcelona, Spain

The concept of harvesting ambient energy as an alternative power source for supplying integrated circuits aiming more miniaturized and distributed applications has been gaining momentum in the past years. A functional energy harvesting system, both in terms of available power and compatibility with system integration, requires concurrently addressing the energy transducing devices together with power management circuits. This talk will address the topic of power management circuits specific for harvesters, particularly emphasizing tight joint characterization, modeling and circuit co-design of the energy transducing devices and the power management frontend integrated circuits, in turn emphasizing vertical co-design for IoT applications.

1 Rationale

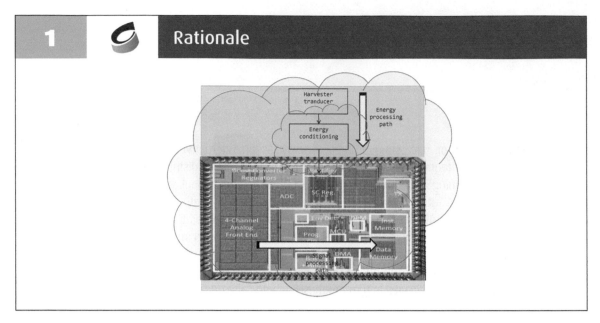

The overarching context of this chapter considers an integrated complex system or SoC, in which there is a signal processing path (encompassing analog, digital and RF signal processing circuits), and orthogonal to that path, and energy processing path, with an energy source (ambient energy), an energy transducer (the harvesting device) and the energy conditioning circuit. While the main performance metric of the former is distortion, resolution or other signal performance descriptions, the latter major metric is efficiency. Since, opposite to power management circuits (DC-DC in particular), the supply path handles time-varying flows, energy is the main magnitude of interest, and as an extension to the field of signal sensors and sensor conditioning circuits, the subsystem in charge of handling efficiently such energy flow is named energy conditioning frontend, which constitutes the main focus of this work.

2 Energy harvesting : the prehistory of a "new" concept

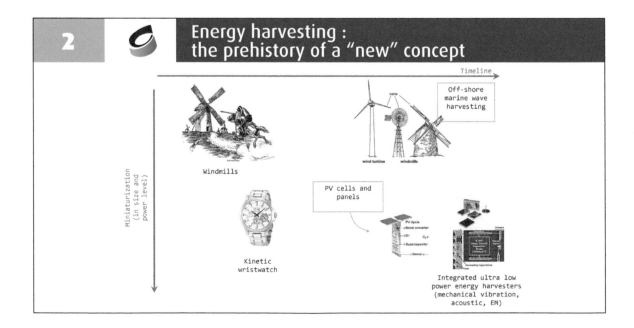

2

Despite the emergence and consolidation of energy harvesting techniques and systems, the very idea of scavenging energy from the ambient can be set in both a timeline and scale of size and energy by identifying what, strictly speaking, constitute instances of harvesting systems and are predecessors of current integrated silicon energy harvesting systems. Indeed, windmills do convert ambient energy in the form of bursts of wind into rotating mechanical energy used for mechanical work. More closely in time, kinetic wristwatches represent another example of converting time-varying bursts of energy coming from the human body into mechanical work. Beyond the downscalability in size, in this last application it is apparent the asynchronicity of the advent of energy and its use, a fundamental property of EH systems, which requires temporary energy storage reservoirs, being in this case a mechanical spring. In this case the energy path is in the mechanical domain, whereas in the case of off-shore marine wave harvesting, the electrical version of windmills, there is a conversation to the electrical domain, which facilitates storage and transportation of energy. Finally, a downscaled version of a system, both in volume/size and energy levels, in which ambient energy is converted, stored, processed and delivered to an (IC) electronic load, constitutes what's the commonplace interpretation of EH systems, particularly those integrated.

3   Energy harvesting : the concept

Energy harvesting **definition**

- Energy harvesting or scavenging consists of **transducing ambient** energy into electrical energy targeting self-powered ICs

  - The qualifying nature of Harvesting Energy sources is their **ubiquitous** and **perpetual** character, but **intermittent/burst** mode

But NOT a *perpetuum mobile*:

In EH the system is NOT closed, but the environment is for ambient energy-abundant (but power scarce)

- This definition naturally leads to a candidate of interest to substitute battery-operated low power equipment in pursuit of **batteryless, self-powered, ever-energy-sustainable**

Accordingly, a working definition of EH needs to be put in place. An EH consists of the transduction of ambient energy into electrical energy, in pursuit of self-powered electronic systems. It is crucial that the energy in the ambient could be of various natures and origins, but what's common out of the diverse and heterogeneous sources of ambient energy, this is, what qualifies a system to be an EH system is the fact the the energy is not punctual, neither in space or time, but rather ubiquitous and perpetual in character. The source of energy, despite distributed and scattered in space has to be everywhere, as an energy field, and its presence should be guaranteed, on average, across time, despite it manifesting in intermittent or burst-mode energy events. These characteristics could describe acoustic energy, thermal gradients, EM waves. Note that a system that detects, extracts and converters energy from the ambient to self-power an electronic -integrated- load is not an oxymoron, despite there was discussion in the scientific community that it would seem a "perpetuum mobile", since the system is not closed energetically but open; the EH system is indeed open to the ambient.

4 Application Motivation

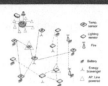

- Wireless sensor networks (WSN) motes
  - On-chip power management for **energy harvesting:**
    - Ultra low power regime
    - High (average) energy efficiency is a must
    - Sophisticated functionality
    - (time-varying burst-mode energy source and load)
    - Multiple heterogeneous energy sources (EM, vibration, acoustic)

- **Biomedical** applications (silicon implants) / RF ID tags
  - Ultra low power, high efficiency critical
  - Variable input energy (inductive link, wireless power transfer)

If it is assumed that EH systems that extract energy from intermittent, ubiquitous, perpetual sources of energy from the ambient to supply electronic loads are feasible, there would be a strong application traction. Certainly, various applications would benefit from the availability of such low-power (but potentially high energy density) sources of energy, namely, among others, (a) wireless sensor node meshes, in which multiple sensors that collaborate wirelessly to sense, exchange and collectively relay signals, do require energy to operate. One of the main bottlenecks that precludes practical deployment of wireless sensor networks, and their scalability, is the availability of energy and the need to avoid replacing batteries if such primary energy sources are considered. A subset of such application would be in the field of biomedical sensors and Body-Area Networks. The type of energy sources in this case would include those inherent to the presence of a human body, in a clinical or ambulatory setting, such as temperature gradients and movement (both regular motion, tremors or even fluid movement internal in the body).

5 Self-Powered or ambient-powered electronics

- Challenge:
  - Using Energy harvesting to provide self-powered nodes

- Energy Harvesting
  - Mechanical Energy: Vibrations or deformations
  - Thermal Energy: Points at different temperatures
  - Solar or light Energy: Sun or artificial light can be harvested with photodiodes
  - Acoustic Energy: Ubiquitous. Energy harvested with MEMS
  - Electromagnetic Energy: Ubiquitous. Energy harvested with *witricity* or antennas.

- All of these power sources can be found in nature and are unlimited.

- In any case, any of theses energy sources are time-variant

There exist a wide variety of energy sources that qualify as sources of energy for EH systems. (a) In the case of mechanical energy, both vibrations or deformations (b) for thermal energy, the gradients of temperature between two distinct spatial points can be converted by leveraging the Seebeck effect in Peltier cells, (c) in the case of solar or light energy, PV cells or photodiodes intrinsically generate electrical energy out of light, (d) for acoustic energy, which is ubiquitous, resonating membranes (or suspended cantilevers) can be used, particularly when miniaturized via MEMs technology implementations, to extract energy from the incident longitudinal waves, a principle analogous to electret microphones, though the design objective here is to maximize efficiency and extracted energy and not signal distortion and noise (e) converting various forms of electromagnetic, RF or magnetic fields to extract power supply electrical energy, including antennas with rectifying RF-DC capabilities or rectennas.

6   Energy Harvesting representative figures

Transduction	Reference	Mechanical input	Output power	Dimensions	Power density
Piezoelectric	Roundy et al. [8] Design 1	f=85 Hz accel=2.25 m/s²	207 uW @ 10V	1cm³	207 uW/cm³
	Roundy et al. [8] Design 2	f=60 Hz accel=2.25 m/s²	335 uW @12 V	1 cm³	335 uW/cm³
	Roundy et al. [8] Design 3	f=40 Hz accel=2.25 m/s²	1700 uW @12 V	4.8 cm³	354 uW/cm³
Magnetic induction	Williams et al. [9]	f=4 kHz Amplitude=300nm	0.3 uW	mm³	300 uW/cm³
	Li et al. [16]	f=64 Hz Amplitude=1mm	10 uW @ 2V	1 cm³	10 uW/cm³
	Yuen et al. [10]	f= 80 Hz Amplitude = 250 um	120 uW @ 900 mV	2.3 cm³	52 uW/cm³
Electrostatic	Meninger et al. [11]	f=2.53 Hz amplitude -	8 uW	0.075 cm3	107 uW/cm³
	Sterken et al [17]	f=1,200 Hz amplitude = 20 um	100 uW @2 V	-	-
	Miyazaki et al. [12]	f=45 Hz amplitude= 1 um	120 uW	-	-

Main characteristics of the millimeter and micrometer scale kinetic energy harvesters

Are those values of harvested energy density enough for current technology?: unclear, but ultralowpower IC techniques anticipate convergence in the immediate coming years.

The question then arises of which are the representative figures and performance metrics attainable for the different technologies for the harvester devices. Being the main performance metric the output extractable energy, and with the quest for miniaturization and the impact of the occupied volume upon the complete system, the most complete and descriptive metric is power density. For the case considered here of kinetic harvester devices down to millimeter and micrometer scale shown in the table, and for the case of piezoelectric transduction, magnetic induction and electrostatic harvesters, such power density is in the range of hundreds of micro Watts per cubic centimeter. The state of the art still doesn't allow to supply representative chips for sensing, ultra low power computing and communication, but there exists a scalability trend in pursuit of increasing the output power density (by a concurrent combination of the fundamental physics of the energy transduction, improving the technological implementation, and the energy processing electronics frontend). Though it is still unclear when the convergence would occur, the progress of minimizing also the power demands for SoC allows to argue that that EH system would self-power complex chips in the near future.

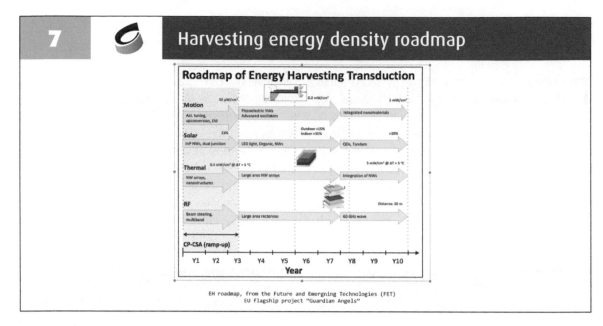

EH roadmap, from the Future and Emergning Technologies (FET)
EU flagship project "Guardian Angels"

To provide quantitative evidence to this statement of converging harvested power and power demands, this slide shows a roadmap for the expected and requested improvement in output energies for various technologies, including motion, solar, thermal and RF, which foresees more than 10x increase in ten years.

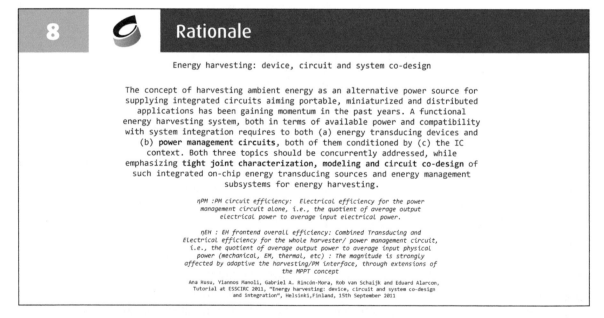

There has recently been a change in focus and need of research and technology in pursuit of complete energy harvesting systems, that beyond the harvester device itself, requires efficient power management circuits, with operation down to the ultra-low-power regime, and both in a integrated circuit context. On top of that, achieving an efficient EH system, given the strong interplay between the different parts, requires a tight joint characterization, modeling and circuit co-design. As an example of that co-design approach, the overall system efficiency, which impacts upon the output available energy for a given energy input profile, depends on both the transduction efficiency and the energy management efficiency.

9 — Challenges for On-chip power management circuits targeting energy harvesting

- A) Further miniaturization of PM circuits aligned to ultra-low-power regimes
- B) Sophisticated functionality (time-varying burst-mode energy source -and load-)
- C) Multiple (heterogenous) energy sources (EM, vibration, acoustic) or scaled harvesters
- D) System-level functionality: Average energy balance is a must with realistic application-driven processing

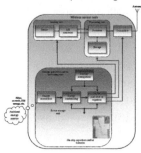

Based on the previous overarching framework and rationale, the conception and design of on-chip power management circuits specifically targeting EH systems faces four crucial challenges, namely (a) the traction to increase the miniaturization levels in both size and volume and power levels, down to the ultra-low-power regimes, in order to co-exist with the harvesting devices and deal with the extremely low power levels as compared to precursor PM circuits used for mobile and low-power power computing in a chip (b) in comparison to DC-DC regulators, EH conditioning does require more sophisticated functionalities given that both the input -and in some applications, the output- is of time-varying nature, usually in the form of bursts (c) energy management circuits for EH need to potentially deal with multiple EH sources and/or multiple scaled harvesters to cope with the distribution of energy in various physical forms and/or frequency bands, and (d) similar to system-driven power management in mobile loads, but with more stringent requirements, Energy management for EH does need to be vertically co-designed, taking into account that the final aim is large timescale average energy balance.

10 — Power MOSFETs

Complete loss optimization of on-chip CMOS synchronous rectifier

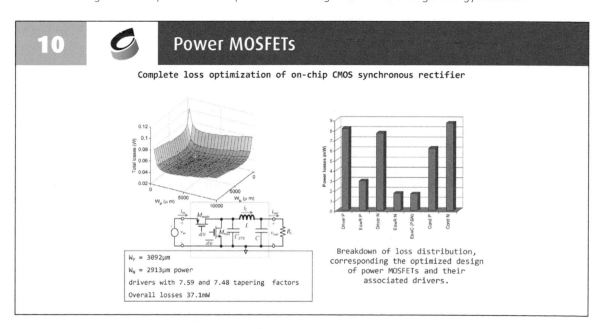

$W_P = 3092\mu m$

$W_N = 2913\mu m$ power

drivers with 7.59 and 7.48 tapering factors

Overall losses 37.1mW

Breakdown of loss distribution, corresponding the optimized design of power MOSFETs and their associated drivers.

In the following section, current techniques and methods for efficient power management are reviewed, with the aim of revisiting them and providing the guidelines for designing energy management for EH. The most basic circuit structure for efficient energy conversion, namely a switch-mode power converter of the step-down buck type, with synchronous rectifier implementation of the power train for on-chip suitability and low voltage operation, is considered. Other alternative topologies are extensions of this fundamental circuit. Once the circuit topology is selected, dimensioning and sizing the circuit requires to balance the different losses (conduction, switching and driving), that usually show a tradeoff upon the width of the transistors, and deciding the tapering factors if tapered buffers are considered to drive the transistor gates. A design-space optimization of the sizes (widths, assuming minimum length, -or slightly larger, to accommodate voltage blocking without compromising area-) of the main power transistors as well as the buffer allows such dimensioning. Selecting the nominal switching frequency stands from the tradeoff of size of the reactive components versus overall efficiency.

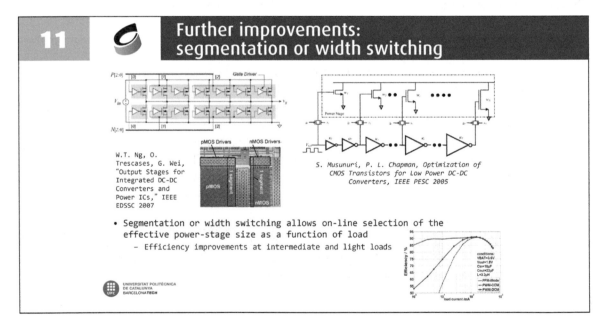

11 Further improvements: segmentation or width switching

W.T. Ng, O. Trescases, G. Wei, "Output Stages for Integrated DC-DC Converters and Power ICs," IEEE EDSSC 2007

S. Musunuri, P. L. Chapman, Optimization of CMOS Transistors for Low Power DC-DC Converters, IEEE PESC 2005

• Segmentation or width switching allows on-line selection of the effective power-stage size as a function of load
  – Efficiency improvements at intermediate and light loads

UNIVERSITAT POLITÈCNICA DE CATALUNYA BARCELONA TECH

In the context of power loads that span a large dynamic range of variations, up to several decades, representative of mobile loads and aggressive operating system level power management, modulation loads include pulse frequency modulation of PFM, which provides a load-dependent variation of the power converter switching converter to equalize the efficiency across a large range of output currents. Analogously, power transistors can be made dynamically varying their equivalent width to cope with varying loads and hence optimally track the balance of switching losses and conduction losses (which depend upon output load). In EH systems, a similar idea can be explored when considering inputs which are time-varying.

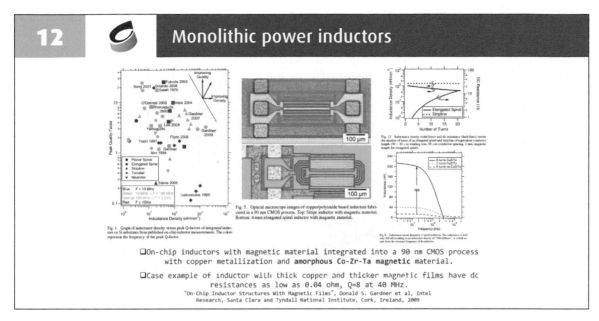

12 **Monolithic power inductors**

Fig. 1. Graph of inductance density versus peak Q-factors of integrated inductors on Si substrates from published on-chip inductor measurements. The colors represent the frequency of the peak Q-factor.

Fig. 5. Optical microscope images of copper/polyimide based inductors fabricated in a 90 nm CMOS process. Top: Stripe inductor with magnetic material. Bottom: 4-turn elongated spiral inductor with magnetic material.

Fig. 13. Inductance density (solid lines) and dc resistance (dark lines) versus the number of turns of an elongated spiral and stripline of equivalent conductor length (50 × 10 μm winding area, 50 μm conductor spacing, 2 mm magnetic length for elongated spiral).

Fig. 8. Inductance versus frequency of spiral inductors. The inductance is well over 200 nH resulting in an inductance density of 1700 nH/mm². A rolloff occurs from the resonant frequency of the inductor.

☐On-chip inductors with magnetic material integrated into a 90 nm CMOS process
with copper metallization and **amorphous Co-Zr-Ta magnetic** material.

☐Case example of inductor with thick copper and thicker magnetic films have dc
resistances as low as 0.04 ohm, Q=8 at 40 MHz.
"On-Chip Inductor Structures With Magnetic Films", Donald S. Gardner et al, Intel
Research, Santa Clara and Tyndall National Institute, Cork, Ireland, 2009

Whean fully monolithic switching power converters are considered, for extreme miniaturization and integration, the main component that precludes such integration is the inductor. Not only they are large reactive devices that occupy area, but they are not immediately amenable to integration. There has been an evident progress in the past decade in technologies for power coil integration. Note that, different to the domain of of RF coils that need to emulate a given narrowband impendence in a frequency band and hence can be sometimes emulated, inductors for power processors operate in baseband with broadband characteristics, and need to physically store -magnetic- energy. This energy storage is crucial to the voltage-current-voltage energy conversion for voltage-voltage adjustable power conversion. State of the art technologies, as shown in this figures, consider integrating magnetic materials and optimized planar geometries to reach hundreds of microhenries per squared millimeter.

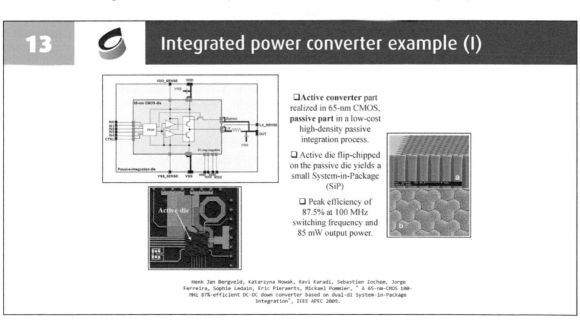

13 **Integrated power converter example (I)**

☐**Active converter** part realized in 65-nm CMOS, **passive part** in a low-cost high-density passive integration process.

☐ Active die flip-chipped on the passive die yields a small System-in-Package (SiP)

☐ Peak efficiency of 87.5% at 100 MHz switching frequency and 85 mW output power.

Henk Jan Bergveld, Katarzyna Nowak, Ravi Karadi, Sebastien Iochem, Jorge
Ferreira, Sophie Ledain, Eric Pieraerts, Mickael Pommier, " A 65-nm-CMOS 100-
MHz 87%-efficient DC-DC down converter based on dual-di System-in-Package
integration", IEEE APEC 2009.

13

Integrating monolithically a power converter might require system-level integration solutions, as this case example shows. Indeed, the power train and the active part is integrated in a 65nm CMOS process, for improved efficiency, compacity, and cost, whereas the reactive components are implemented in a high-density passive integration process including trench capacitors an thick metal plane high quality inductors. The final integrated system achieves, through active die flip-chip co-integrated in a dedicated System-in-Package solution, a performance which this particular case is able to deliver 85 mV of power at 87.5% peak efficiency by switching at 100 MHz.

14 Integrated power converter example (II)

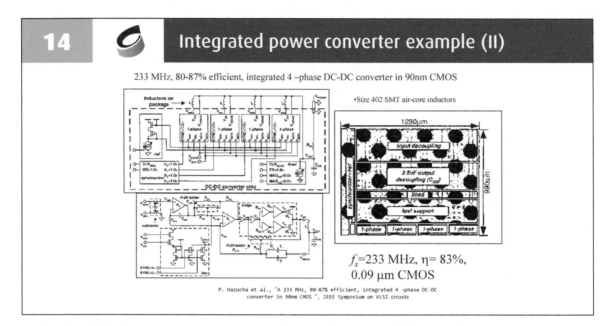

233 MHz, 80-87% efficient, integrated 4 –phase DC-DC converter in 90nm CMOS

f_s=233 MHz, η= 83%, 0.09 µm CMOS

P. Hazucha et al., "A 233 MHz, 80-87% efficient, integrated 4 –phase DC-DC converter in 90nm CMOS ", IEEE Symposium on VLSI circuits

A more aggressive solution to push integration and miniaturization limits considers a 90nm CMOS process with switching frequencies exceeding 200 MHz and an interleaved multi-converter architecture. Synchronous rectification and tapered buffers in CMOS together with self-oscillating hysteric control for each phase, with a four-phase combination enabled by off-chip in-package air-core inductors results in efficiencies in the 80%-87% range. Both output filter capacitor as well as input decoupling capacitor (given the switching characteristics of the input current of a buck converter stage, that could results in voltage modulation at the input voltage accounting for non-negligible resistance of input pads and bonding wires) are co-integrated planarly. This solution aims integrated conversion for mobile microprocessors, and sets the limit of miniaturization given the extreme high frequencies.

15 · Integrated power converter example: SIMO LC buck

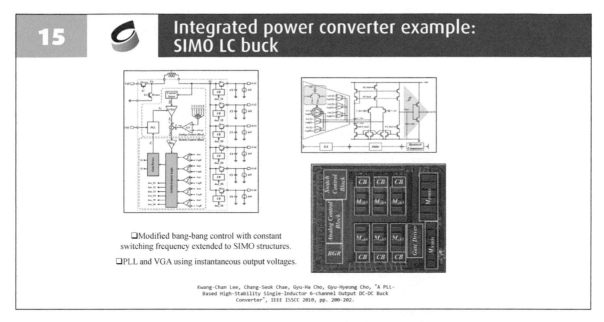

❑ Modified bang-bang control with constant
switching frequency extended to SIMO structures.

❑ PLL and VGA using instantaneous output voltages.

Kwang-Chan Lee, Chang-Seok Chae, Gyu-Ha Cho, Gyu-Hyeong Cho, "A PLL-
Based High-Stability Single-Inductor 6-channel Output DC-DC Buck
Converter", IEEE ISSCC 2010, pp. 200-202.

Architectures of power management exist that enable supplying multiple outputs, and they representative of low power systems that need different voltage and current levels at different output ports. In this case, single inductor converters, by appropriate time-multiplexing switching policies, energize the single inductor and implement power steering to supply the different outputs. The ease of integration thanks to using a single inductor can be maintained in EH systems with multiple inputs, thereby implementing multiple-input multiple-output MIMO single-inductor switching power converters.

16 · Complete integrated 3-level CMOS switching power converter

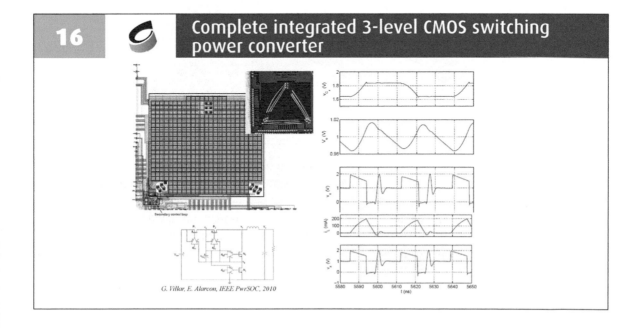

G. Villar, E. Alarcon, IEEE PwrSOC, 2010

A representative example of a fully monolithically integrated switching power converter in standard CMOS is discussed here. Despite this integration technology is not naturally suited to integrating the power train and the power reactives, still some decent performance can be obtained by optimizing the structure/method, the geometry and the design of the different components. Indeed, in this case planar power MOSFETs with optimized balance of losses in the switching loss/conduction loss/driver loss space are considered. Output power capacitors for filtering consider aiming for minimazing Equivalent Series Resistance (ESR) by using a matrix of aspect-ratio-optimized MOSFET capacitors (channel capacitance with appropriate loss balancing of gate resistance and channel resistance, by leveraging the high density –nonlinear– capacitance of the channel, the voltage-dependence characteristics of which are not critical in a DC-DC operation) designed to provide a bounded series impedance at a given operation frequency. On-chip/on-chip triangular bonding wire inductors result in (a) available area underneath the structure as well as (b) low resistance in CMOS. Finally, a floating capacitor stacked-transistor 3-level multiple phase switching topology trades-off output ripple, efficiency and efficiency.

17 Optimized design space exploration

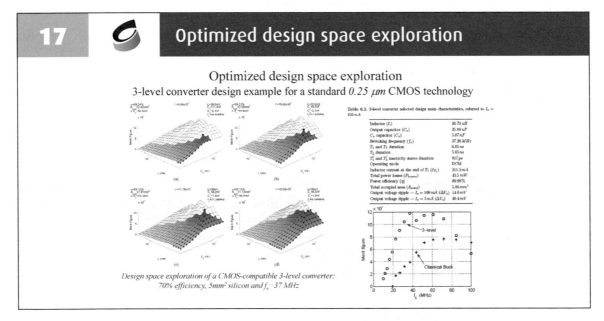

Design space exploration of a CMOS-compatible 3-level converter:
70% efficiency, 5mm² silicon and f_s =37 MHz

Following with the description of the standard CMOS 3-level DC-DC converter, despite using optimized approaches for the reactive components, the power train and the converter topology, still the challenge exists of providing an optimized parametric design of the converter, the design space composed by the switching frequency, L and C. A structured approach design methodology based upon an optimized design space exploration defining as performance metrics the efficiency, the occupied silicon area, and the output ripple acting as a constrain or threshold, combined in a single quantitative figure of merit representing the overall performance goodness of the design allows, by the appropriate characterization- and simulation-based models, provide as outcomes the optimal design parameters. Beyond that, this design optimization framework allows to provide a benchmark comparison with alternative cases, which in this case are the optimized 3-level converter compared to an equivalent 2-level converter, showing, in a comparison framework with a homogeneous comparison metric (overall merit figure) and the same input design parameter – the switching frequency– an enhanced performance which outperforms in a factor of more than a decade.

18 Integrated circuit examples: SC (I)

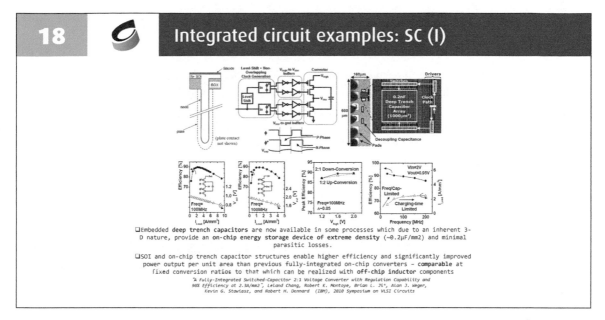

❑Embedded **deep trench capacitors** are now available in some processes which due to an inherent 3-D nature, provide an **on-chip energy storage device of extreme density** (~0.2µF/mm2) and minimal parasitic losses.

❑SOI and on-chip trench capacitor structures enable higher efficiency and significantly improved power output per unit area than previous fully-integrated on-chip converters - **comparable at** fixed conversion ratios to that which can be realized with **off-chip inductor** components

"A Fully-Integrated Switched-Capacitor 2:1 Voltage Converter with Regulation Capability and 90% Efficiency at 2.3A/mm2", Leland Chang, Robert K. Montoye, Brian L. Ji, Alan J. Weger, Kevin G. Stawiasz, and Robert H. Dennard (IBM), 2010 Symposium on VLSI Circuits

Alternatives to inductive switching power converters exist which are based on -power-switches and high quality -power- capacitors to provide DC-DC conversion. The fundamental operating principle of these converters, which are fixed in DC-DC conversion ration by topology, is inspired by high voltage generators and voltage doublers for which step-up doubling is obtained by periodically connecting in parallel (for charging) and in series (to step up voltage) capacitors. Based on this principle and depending on the topology, fixed DC-DC conversion ratios can be implemented. The main advantage is suitability of on-chip integration (no inductors). The main performance bottleneck is the series resistance of capacitors. Hence, this example shows using high quality capacitors by using trench capacitors in memory processes with trench technologies. The low series resistance of this capacitor technology results in, in this example from IBM, operating efficiencies of up to 90% for switching frequencies of 100 MHz and high density operation in the order of various Amps per square millimeter.

19 Integrated circuit examples: SC (II)

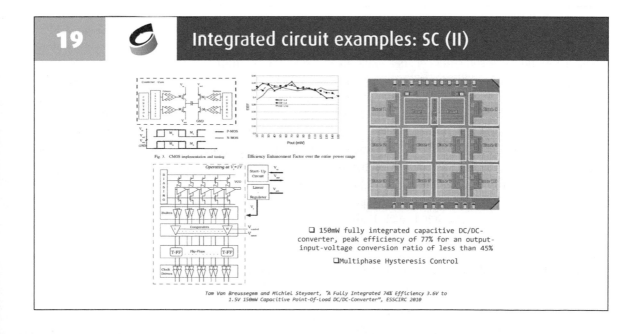

❑ 150mW fully integrated capacitive DC/DC-converter, peak efficiency of 77% for an output-input-voltage conversion ratio of less than 45%

❑Multiphase Hysteresis Control

Tom Van Breussegem and Michiel Steyaert, "A Fully Integrated 74% Efficiency 3.6V to 1.5V 150mW Capacitive Point-Of-Load DC/DC-Converter", ESSCIRC 2010

19

If standard CMOS technologies are considered to implement switch-capacitor DC-DC converters, the intrinsic impairments in the quality of planar capacitors and their loss-inducing parasitic capacitors can be circumvented by more elaborated circuit topologies and system architectures. In this representative case of a high-performance switch-capacitor DC-DC converter in standard CMOS, which performs with up to 77% efficiency for peak power levels of 150mW (a metric that should better be compared with what an LDO counterpart could achieve for a given voltage ratio, better than with the deviation from the ideal 100%). In this case example a matrix-like regular architecture based on unitary cells achieves fine-grain DC-DC regulation by a programmable distributed gear-box approach.

20 ## Technology state of the art of integrated power management ICs

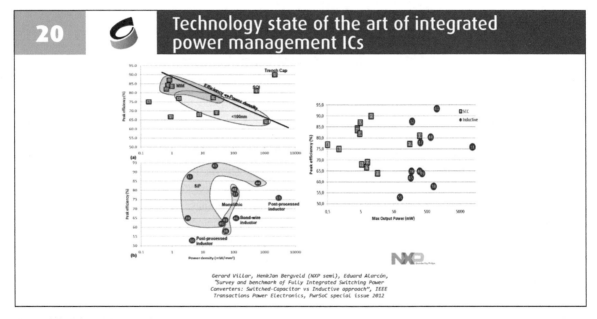

Gerard Villar, HenkJan Bergveld (NXP semi), Eduard Alarcón,
"Survey and benchmark of Fully Integrated Switching Power
Converters: Switched-Capacitor vs Inductive approach", IEEE
Transactions Power Electronics, PwrSoC special issue 2012

The question arises, particularly when aiming for EH systems in the ultra-low-power regime, of whether the inductive converters or their capacitive counterparts are more suitable for such scenario. Here it is discussed a comparison of both approaches for the state of art in terms of performance metrics, which also given insights on future trends. In the first upper left plot, the description of the achievable peak efficiency as a function of the converter power density (in mmW per square mm in planar implementations) yields that for switched-capacitor converters there is a main trend that tradeoffs efficiency versus power efficiency, with improved performance in MIM capacitor technologies for lower power levels, and additionally Silicon-over-insulator technologies and even Trench capacitors are able to outperform this main trend, yielding examples up to 90% for 1W/mm2. For the inductive case, System-in-Package approaches yield a cluster of cases in the 10-100 mW /mm2 range with efficiencies between 70% and 93%, with monolithic approaches being adequate for power densities around 100mW/mm2 with more moderate efficiencies, and post-processed inductors resulting in moderate efficiencies up to 75% for densities larger than 1W/mm2. Finally, a relative comparison shows that inductive approaches are naturally suited for higher output power levels, with SC naturally operating in lower power levels. It is then argued that EH should upfront consider SC approaches.

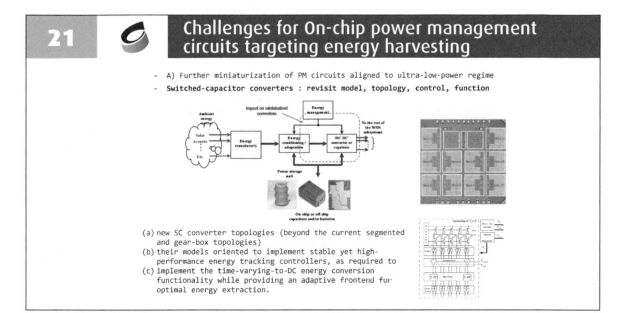

21 Challenges for On-chip power management circuits targeting energy harvesting

- A) Further miniaturization of PM circuits aligned to ultra-low-power regime
- Switched-capacitor converters : revisit model, topology, control, function

(a) new SC converter topologies (beyond the current segmented and gear-box topologies)
(b) their models oriented to implement stable yet high-performance energy tracking controllers, as required to
(c) implement the time-varying-to-DC energy conversion functionality while providing an adaptive frontend for optimal energy extraction.

Once realized that potentially SC power converters suit better EH applications operating in the ultra-low-power regime, it is important to state the main challenges that are still open when considering this power converter circuits to provide energy conditioning in energy harvesting applications. Further miniaturization of PM circuits aligned to ultra-low-power regime, as required by the application, requires for switch-capacitor converters to revisit their topologies and functions, their models and their control methods and circuits. In particular, new SC converter topologies, beyond the current segmented and gear-box topologies, need to be explored to cope with non DC-DC functions, as required in EH systems with time-varying inputs and outputs. There is the need to implement the time-varying-to-DC energy conversion functionality while providing an adaptive frontend for optimal energy extraction. Finally, their behavioral dynamic models oriented to implement stable yet high-performance energy tracking controllers should be further explored.

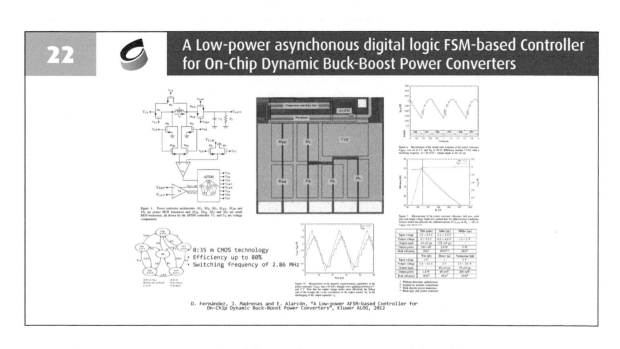

22 A Low-power asynchonous digital logic FSM-based Controller for On-Chip Dynamic Buck-Boost Power Converters

- 0:35 m CMOS technology
- Efficiency up to 80%
- Switching frequency of 2.86 MHz

D. Fernández, J. Madrenas and E. Alarcón, "A Low-power AFSM-based Controller for On-Chip Dynamic Buck-Boost Power Converters", Kluwer ALOG, 2012

When considering switch-mode controllers, which include both the compensation and control itself and then the modulation to convert the feedback signal into a pulse switching pattern, for either inductive or capacitive power converters, the conventional constant switching frequency PWM control, its extension to PFM, or nonlinear switching controllers as hysteretic control are usual when dealing with DC-DC regulation. In the context of time-varying power inputs and outputs, other approaches can be considered. In this example, a low-power controller based upon a finite-state-machine digital control that is not generating a high frequency modulation, but instead a zone-based control is considered for a fully integrated non-inverting buck-boost inductive converter for tracking applications.

23 Enhanced switching policy for deep multi-level power conversion

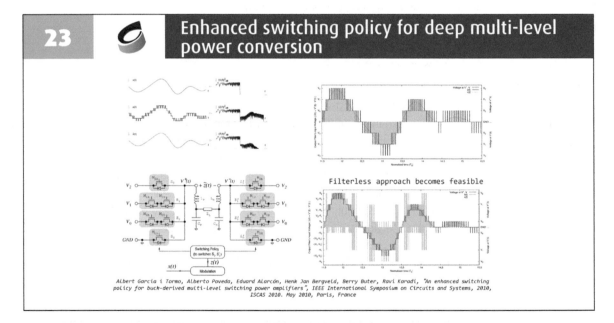

Filterless approach becomes feasible

Albert Garcia i Tormo, Alberto Poveda, Eduard Alarcón, Henk Jan Bergveld, Berry Buter, Ravi Karadi, "An enhanced switching policy for buck-derived multi-level switching power amplifiers", IEEE International Symposium on Circuits and Systems, 2010, ISCAS 2010. May 2010, Paris, France

When time-varying outputs are considered in power amplifiers or time-varying power supplies, the concept of multiple levels can be taken to the limit so that a filterless approach becomes feasible. This case could be considered in the EH systems to treat directly the input time-varying power waveforms, providing the appropriate conversation but without the need of power reactives, thereby potentially facilitating on-chip integration.

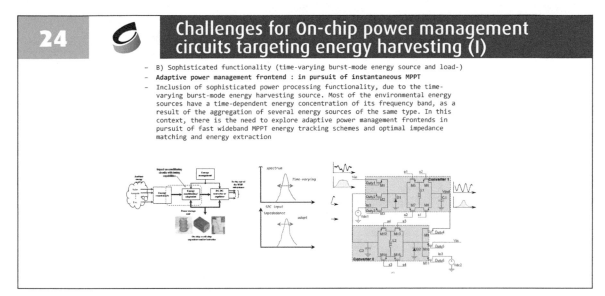

24 — Challenges for On-chip power management circuits targeting energy harvesting (I)

- B) Sophisticated functionality (time-varying burst-mode energy source and load-)
- **Adaptive power management frontend : in pursuit of instantaneous MPPT**
- Inclusion of sophisticated power processing functionality, due to the time-varying burst-mode energy harvesting source. Most of the environmental energy sources have a time-dependent energy concentration of its frequency band, as a result of the aggregation of several energy sources of the same type. In this context, there is the need to explore adaptive power management frontends in pursuit of fast wideband MPPT energy tracking schemes and optimal impedance matching and energy extraction

Other challenges for on-chip power management circuits targeting energy harvesting consist in the inclusion of sophisticated power processing functionality, due to the time-varying burst-mode energy harvesting source. Most of the environmental energy sources have a time-dependent energy concentration of its frequency band, as a result of the aggregation of several energy sources of the same type. In this context, there is the need to explore adaptive power management frontends in pursuit of fast wideband MPPT energy tracking schemes and optimal impedance matching and energy extraction.

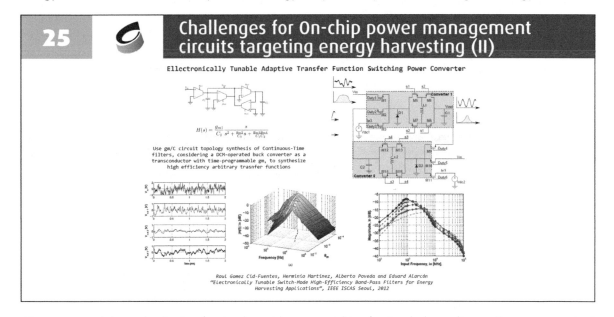

25 — Challenges for On-chip power management circuits targeting energy harvesting (II)

Ellectronically Tunable Adaptive Transfer Function Switching Power Converter

$$H(s) = \frac{g_{m1}}{C_1} \frac{s}{s^2 + \frac{g_{m2}}{C_1}s + \frac{g_{m3}g_{m4}}{C_1 C_2}}$$

Use gm/C circuit topology synthesis of Continuous-Time filters, considering a DCM-operated buck converter as a transconductor with time-programmable gm, to synthesize high efficiency arbitrary trasnfer functions

Raul Gomez Cid-Fuentes, Herminio Martinez, Alberto Poveda and Eduard Alarcón
"Electronically Tunable Switch-Mode High-Efficiency Band-Pass Filters for Energy Harvesting Applications", IEEE ISCAS Seoul, 2012

One approach for such adaptive frontend considers synthesizing switch-mode efficient power converter topologies that are electronically tunable, in this example illustrated to implement tunable bandpass switching circuits that can be extended to implement electronically tunable impedance emulation circuits for instantaneous fast impedance matching frontends depending on the energy content of the input energy. In this case, it is considered the use of gm/C circuit topology synthesis of Continuous-Time filters, considering a DCM-operated buck converter as a transconductor with time-programmable (average) gm, to synthesize high efficiency switch-mode tunable arbitrary transfer functions.

26

Challenges for On-chip power management circuits targeting energy harvesting (III)

- **System/circuit/device co-design (II)**
  - Co-design power converter and Harvester (embed functionality)

Frontend PM circuit designs conceived to optimally extract harvested energy while simplifying circuit redundancy require breaking down simple block diagrams and using a circuit/device co-design approach in the form of jointly co-designing the power converter and the harvester device

Block diagram is too simplistic in Energy harvesting applications: need to model and control bidirectional energy transfers and impedances: matrix impedance models

-Different to low power analog: **impedances** for energy match are key

-Different to RF: power **bidirectionality** is key

A crucial aspect of energy conditioning circuit design, qualitatively different from power management circuits, is the need to co-design the power converter together with the harvester device, by embedding functionalities. Indeed, frontend PM circuit designs conceived to optimally extract harvested energy while simplifying circuit redundancy require breaking down simple block diagrams and using a circuit/device co-design approach in the form of jointly co-designing the power converter and the harvester device. The reason for this is that block diagrams are too simplistic in Energy harvesting applications, since although there is the need to describe a multi-port system, there is the the need as well to model and control bidirectional energy transfers and impedances, the latter including matrix impedance models. To highlight the relevance of this, it is instructive to realize that, different to low power analog circuit design, in EH systems impedance matching for energy transfer is key, and, different to RF circuit architectures, energy bidirectionality can be relevant in EH energy management frontends.

27

Harvester-circuit co-design (I)

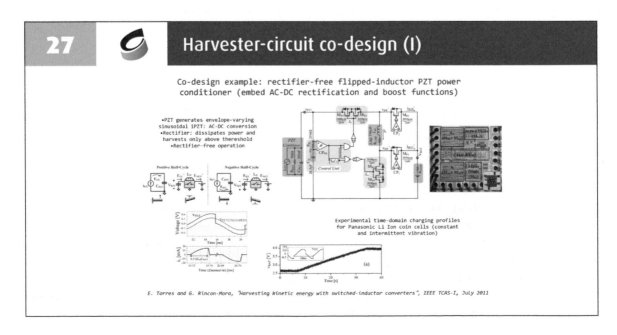

Co-design example: rectifier-free flipped-inductor PZT power conditioner (embed AC-DC rectification and boost functions)

- PZT generates envelope-varying sinusoidal iPZT: AC-DC conversion
- Rectifier: dissipates power and harvests only above thereshold
- Rectifier-free operation

Positive Half-Cycle   Negative Half-Cycle

Experimental time-domain charging profiles for Panasonic Li Ion coin cells (constant and intermittent vibration)

E. Torres and G. Rincon-Mora, "Harvesting kinetic energy with switched-inductor converters", IEEE TCAS-I, July 2011

27

An example of a harvester/energy management circuit co-design approach is presented here. The circuit, instead of serially combining the different functionalities needed to condition the power of a piezoelectric energy source (energy AC-DC rectification and DC-DC step-up functions due to low energy -and accordingly voltage- levels), considers a single stage that embeds both functionalities in a rectifier-free flipped-inductor power conditioner. The validation of the compact integrated circuits shows how the PZT generates envelope-varying quasi-sinusoidal energy that requires AC-DC conversion. The rectifier dissipates power and harvests only above threshold. The rectifier-free operation by virtue of a flipped inductor non-inverting buck-boost operation provides the overall functionality. Experimental time-domain results show charging profiles for Li Ion coin cells under both constant and intermittent vibration excitation.

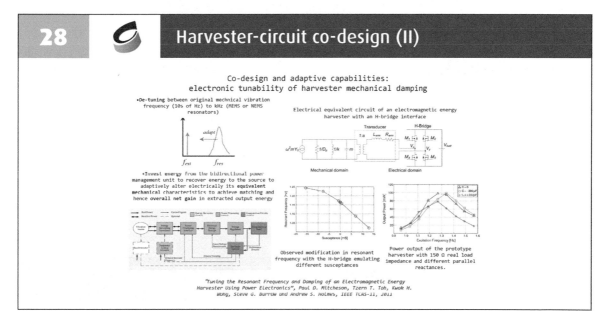

28 **Harvester-circuit co-design (II)**

Co-design and adaptive capabilities:
electronic tunability of harvester mechanical damping

Observed modification in resonant frequency with the H-bridge emulating different susceptances

Power output of the prototype harvester with 150 Ω real load impedance and different parallel reactances.

"Tuning the Resonant Frequency and Damping of an Electromagnetic Energy Harvester Using Power Electronics", Paul D. Mitcheson, Tzern T. Toh, Kwok H. Wong, Steve G. Burrow and Andrew S. Holmes, IEEE TCAS-II, 2011

Another example of a co-design with adaptive capabilities considers the electronic tunability of mechanical damping for a MEMs mechanical resonator. Indeed, in these harvesters a crucial impairment consists of the fundamental de-tuning between the original mechanical vibration frequency (10s of Hz) to the natural resonant frequency (in the order of kHz for miniaturized MEMS or NEMS resonators). This fact precludes operation of the system, yielding no harvested energy. A possible solution that considers a smart energy management frontend that leverages bidirectional energy flow, considers –counterintuitively– investing energy from the bidirectional power management that it is injected back to the harvester unit to recover energy to the source to adaptively alter electrically its equivalent mechanical characteristics to achieve matching and hence an overall net gain in extracted output energy. The figures show the electrical equivalent circuit of the electromagnetic energy harvester with an H-bridge interface, and the experimental validation curves provide evidence of the observed modification in the resonant frequency with the H-bridge emulating different susceptances.

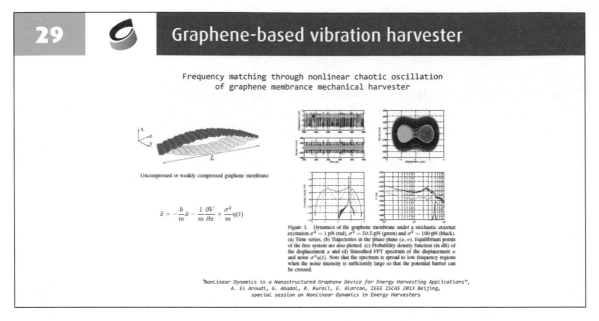

"Nonlinear Dynamics in a Nanostructured Graphene Device for Energy Harvesting Applications",
A. El Aroudi, G. Abadal, R. Rurali, E. Alarcon, IEEE ISCAS 2013 Beijing,
special session on Nonlinear Dynamics in Energy Harvesters

A final example of a relevant approach on how to deal with the fundamental frequency mismatch between input excitation frequencies and the resonant frequencies of resonators considers the nonlinear vibration of the suspended membrane, the nonlinear dynamics of which can be designed to sustain subharmonic or chaotic modes that result in spectral broadening that downconverters the transduction process. In this particular -more exotic- example, the extreme case of miniaturized membranes down to the nanoscale, leverage the properties of graphene, a material that results in bistable dynamic behaviour of the the graphene-based vibration harvester under lateral mechanical compression. The frequency matching through nonlinear chaotic oscillation of the graphene membrane mechanical harvester is due to the change of mechanical properties of the compressed graphene membrane.

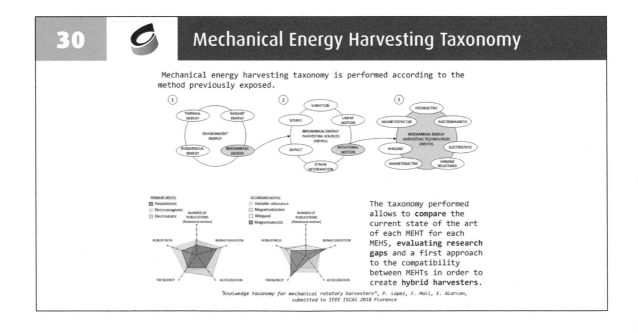

"Knowledge taxonomy for mechanical rotatory harvesters", P. Lopez, F. Moll, E. Alarcon,
submitted to IEEE ISCAS 2018 Florence

30

It is argued here that the field of energy harvesters has been developing in the past years so that there is a timely need to structure the generated knowledge (techniques, devices, circuits) by a taxonomy. In this case, the particular subcase of mechanical energy harvesting is discussed. Such taxonomy allows (a) the categorization and structuring of the different alternatives given the specification of a given application (b) inverting such classification into a design-oriented selection and dimensioning tool, and even (c) identifying the research and technology gaps as well as unexplored combinations that could result in hybrid harvesters with improved performance.

About the Editors

Mathieu Coustans

M athieu Coustans received a M.S. degree in Electrical Engineering from INPT-ENSEEIHT in 2014.

He is currently with the Electronics Laboratory at the Swiss Federal Institute of Technology (EPFL) and is working towards a PhD degree.

His work focuses on ultra-low power/energy-efficient timing and energy harvesting Integrated Circuits.

His research interests include device reliability and modelling, CMOS analog and mixed-signal energy-efficient circuit design.

Catherine Dehollain

M y name is Catherine Dehollain, I am Professor in electronics at EPFL and I am responsible for the RFIC group. One of the main research topics of our group is dedicated to remote powering and data communication for sensor networks. We have started to work in this domain since year 2000 and we have got financial support thanks to the *Ecole Polytechnique Fédérale de Lausanne* (EPFL), to the Swiss National Research Foundation (SNF) for basic and fundamental research, to the Swiss CTI Office for applied research in collaboration with the Swiss industry as well as to the European commission for applied research in collaboration with other European universities, European research centers and European companies.

About the Authors

Eduard Alarcón

Eduard Alarcón received the M. Sc. (National award) and Ph.D. degrees (honors) in Electrical Engineering from the Technical University of Catalunya (UPC BarcelonaTech), Spain, in 1995 and 2000, respectively. Since 1995 he has been with the Department of Electronics Engineering at the School of Telecommunications at UPC, where he became Associate Professor in 2000. From August 2003 to January 2004, July-August 2006 and July-August 2010 he was a Visiting Professor at the CoPEC center, University of Colorado at Boulder, US, and during January-June 2011 he was Visiting Professor at the School of ICT/Integrated Devices and Circuits, Royal Institute of Technology (KTH), Stockholm, Sweden. During the period 2006-2009 he was Associate Dean of International Affairs at the School of Telecommunications Engineering, UPC. He has co-authored more than 400 scientific publications, 7 books, 8 book chapters and 12 patents, and has been involved in different National, European (H2020 FET-Open, Flag-ERA) and US (DARPA, NSF) R&D projects within his research interests including the areas of on-chip energy management and RF circuits, energy harvesting and wireless energy transfer, nanosatellites, and nanotechnology-enabled wireless communications. He has received the Google Faculty Research Award (2013), Samsung Advanced Institute of Technology Global Research Program gift (2012), and Intel Honor Programme Fellowship (2014). He has given 30 invited, keynote and plenary lectures and tutorials in Europe, America, Asia and Oceania, was appointed by the IEEE CAS society as distinguished lecturer for 2009-2010 and lectures yearly MEAD courses at EPFL. He is elected member of the IEEE CAS Board of Governors (2010-2013), member of the IEEE CAS long term strategy committee, Vice President Finance of IEEE CAS (2015) and Vice President for Technical Activities of IEEE CAS (2016-2017, and 2017-2018). He was recipient of the Myril B. Reed Best Paper Award at the 1998 IEEE Midwest Symposium on Circuits and Systems. He was the invited co-editor of a special issue of the Analog Integrated Circuits and Signal Processing journal devoted to current-mode circuit techniques, a special issue of the International Journal on Circuit Theory and Applications, invited associate editor for a IEEE TPELS special issue on PwrSOC. He co-organized special sessions related to on-chip power management at IEEE ISCAS03, IEEE ISCAS06 and NOLTA 2012, and lectured tutorials at IEEE ISCAS09, ESSCIRC 2011, IEEE VLSI-DAT 2012 and APCCAS 2012. He was the 2007 Chair of the IEEE Circuits and Systems Society Technical Committee on Power Circuits. He is acting as general co-chair of DCIS 2017, Barcelona and IEEE ISCAS 2020, Seville. He was the General co-chair of the 2014 international CDIO conference, the technical program co-chair of the 2007 European Conference on Circuit Theory and Design - ECCTD07 and of LASCAS 2013, Special Sessions co-chair at IEEE ISCAS 2013. He served as an Associate Editor of the IEEE Transactions on Circuits and Systems - II: Express briefs (2006-2007) and Associate Editor of the Transactions on Circuits and Systems – I: Regular papers (2006-2012) and currently serves as Associate Editor Elsevier's Nano Communication Networks journal (2009-), Journal of Low Power Electronics (JOLPE) (2011-) and in the Senior founding Editorial Board of the IEEE Journal on IEEE Journal on Emerging topics in Circuits and Systems, of which he is currently Editor-in-Chief (2018).

[Chapter 10]

Andreas Burg

Andreas Burg (S'97–M'05) was born in Munich, Germany, in 1975. He received the Dipl.Ing. degree from the Swiss Federal Institute of Technology (ETH) Zurich, Zurich, Switzerland, in 2000, and the Dr.Sc.Techn. degree from the Integrated Systems Laboratory, ETH Zurich, in 2006.In 1998, he was with Siemens Semiconductors, San Jose, CA, USA. During his Ph.D. studies, he was with the Bell Labs Wireless Research for a total of one year. From 2006 to 2007, he was a Post- Doctoral Researcher with the Integrated Systems Laboratory and with the Communication Theory Group, ETH Zurich. In 2007, he co-founded Celestrius, an ETH-spinoff in the field of MIMO wireless communication, where he was responsible for the ASIC development as the Director for VLSI. In 2009, he joined ETH Zurich as an SNF Assistant Professor and as the Head of the Signal Processing Circuits and Systems Group with the Integrated Systems Laboratory. Since 2011, he has been a Tenure Track Assistant Professor with the *École Polytechnique Fédérale de Lausanne*, where he is leading the Telecommunications Circuits Laboratory.

[Chapter 5]

Sandro Carrara

S andro Carrara is an IEEE Fellow for his outstanding record of accomplishments in the field of design of nanoscale biological CMOS sensors. He is also the recipient of the IEEE Sensors Council Technical Achievement Award in 2016 for his leadership in the emerging area of co-design in Bio/Nano/CMOS interfaces. He is a faculty member (MER) at the EPFL in Lausanne (Switzerland). He is former professor of optical and electrical biosensors at the Department of Electrical Engineering and Biophysics (DIBE) of the University of Genoa (Italy) and former professor of nanobiotechnology at the University of Bologna (Italy). He holds a PhD in Biochemistry & Biophysics from University of Padua (Italy), a Master degree in Physics from University of Genoa (Italy), and a diploma in Electronics from National Institute of Technology in Albenga (Italy). His scientific interests are on electrical phenomena of nano-bio-structured films, and include CMOS design of biochips based on proteins and DNA. Along his carrier, he published 7 books, one as author with Springer on Bio/CMOS interfaces and, more recently, a Handbook of Bioelectronics with Cambridge University Press. He has more than 250 scientific publications and is author of 13 patents. He is now Editor-in-Chief of the IEEE Sensors Journal, the largest journal among 180 IEEE publications; he is also founder and Editor-in-Chief of the journal BioNanoScience by Springer, and Associate Editor of IEEE Transactions on Biomedical Circuits and Systems. He is a member of the IEEE Sensors Council and his Executive Committee. He was a member of the Board of Governors (BoG) of the IEEE Circuits And Systems Society (CASS). He has been appointed as IEEE Sensors Council Distinguished Lecturer for the years 2017-2019, and CASS Distinguished Lecturer for the years 2013-2014. His work received several international recognitions: several Top-25 Hottest-Articles (2004, 2005, 2008, 2009, and two times in 2012) published in highly ranked international journals such as Biosensors and Bioelectronics, Sensors and Actuators B, IEEE Sensors journal, and Thin Solid Films; a NATO Advanced Research Award in 1996 for the original contribution to the physics of single-electron conductivity in nano-particles; five Best Paper Awards at the Conferences IEEE NGCAS in 2017 (Genoa), MOBIHEALTH in 2016 (Milan), IEEE PRIME in 2015 (Glasgow), in 2010 (Berlin), and in 2009 (Cork); three Best Poster Awards at the EMBEC Conference in 2017 (Tampere, Finland), Nanotera workshop in 2011 (Bern), and NanoEurope Symposium in 2009 (Rapperswil). He also received the Best Referees Award from the journal Biosensor and Bioelectronics in 2006. From 1997 to 2000, he was a member of an international committee at the ELETTRA Synchrotron in Trieste. From 2000 to 2003, he was scientific leader of a National Research Program (PNR) in the filed of Nanobiotechnology. He was an internationally esteemed expert of the evaluation panel of the Academy of Finland in a research program for the years 2010-2013. He has been the General Chairman of the Conference IEEE BioCAS 2014, the premier worldwide international conference in the area of circuits and systems for biomedical applications

[Chapter 1]

Mathieu Coustans

Mathieu Coustans received a M.S. degree in Electrical Engineering from INPT-ENSEEIHT in 2014.

He is currently with the Electronics Laboratory at the Swiss Federal Institute of Technology (EPFL) and is working towards a PhD degree.

His work focuses on ultra-low power/energy-efficient timing and energy harvesting Integrated Circuits.

His research interests include device reliability and modelling, CMOS analog and mixed-signal energy-efficient circuit design.

[Chapter 8]

Catherine Dehollain

My name is Catherine Dehollain, I am Professor in electronics at EPFL and I am responsible for the RFIC group. One of the main research topics of our group is dedicated to remote powering and data communication for sensor networks. We have started to work in this domain since year 2000 and we have got financial support thanks to the *Ecole Polytechnique Fédérale de Lausanne* (EPFL), to the Swiss National Research Foundation (SNF) for basic and fundamental research, to the Swiss CTI Office for applied research in collaboration with the Swiss industry as well as to the European commission for applied research in collaboration with other European universities, European research centers and European companies.

[Chapter 2]

Roberto La Rosa

Roberto La Rosa is currently working as Design Manager and Smart Energy Applications Team Manager for STMicroelectronics Catania. Since Joining STMicroelectronics in 1997 he has held a variety of assignments, including the design of high-frequency PLL's for clock generation and recovery, fiber-optic transceiver and system design, power managements IC, and other analog, digital and mixed-signal bipolar and CMOS circuit development projects. He currently is a Research Senior Staff Member at STMicroelectronics Catania. His current research interests include Ultra low power management, over the distance power transmission and Energy Harvesting.

Dr. La Rosa has published several papers on advanced techniques to null stand-by power consumption by using energy harvesting and holds several patents.

[Chapter 3]

Sherryl Manalo

S herryl has a Ph.D. in theoretical physics and worked in research before switching over to IT. She was the co-founder of a small start-up in Austria with the focus of providing infrastructure services and software development in the field of Linux Server and Open Source Software, operating 6 years in this field. She then joined Microsoft starting in MCS 2008 as Technical Consultant for Windows Server, Client and System Center and soon became Engagement Manager in the service line Core IO, Datacenter Services. As a next step, she was the Account Technology Strategist responsible for Infrastructure Optimization. After relocating to Switzerland, she became TS Practice Lead in Switzerland for SoftwareONE. She then went back to Microsoft beginning 2016 as Cloud Solution Architect to focus on her interest in Azure, specifically IoT scenarios and Machine Learning, which has become the most interesting Cloud platform currently existing in IT.

[Presenter "Microsoft's approach to IoT"]

Marcel Meli

Prof. Marcel Meli (PhD, Cardiff, Great Britain) is lecturer for Computer Engineering and IoT.

He is also the Head of the Wireless System Group at the Institute of Embedded Systems of the University of applied Sciences in Winterthur, Switzerland. His research interests are in the area of low power computing and communication, energy harvesting, power management.

[Chapter 9]

Maria-Alexandra Paun

My name is Dr. Maria-Alexandra Paun and I am currently a Scientist at the Swiss Federal Institute of Technology in Lausanne, Switzerland. I am also the Chair of IEEE Switzerland Section and Chair of IEEE Switzerland Women in Engineering. It will be my pleasure to be delivering today my presentation entitled "Low power Hall effect sensors, in the framework of the doctoral course MICRO-622. From design optimization to CMOS integration". This is original work that has been developed during more than 7 years in the field of Hall effect sensors, including work during my PhD at EPFL, Switzerland (2009-2013) and postdoctoral research fellowships at University of Cambridge, UK (2013-2015) and published in various articles.

[Chapter 6]

Pascal Urard

Pascal Urard graduated in 1991 from ISEN, Lille, France.

He joined SGS-Thomson in 1992 where he has been involved successively in test and engineering of mixed signal ASICs, digital ASIC design, and architecture of signal processing ASICs in the domain of digital communications. He joined STMicroelectronics research center, Crolles, France, in 2000 to work on advanced design flows. In 2010, he initiated the first autonomous IPv6 wireless network for sensors and actuators (GreenNet), demonstrating bidirectional secured IPv6 communications over the air powered by energy harvesters. As an ST Fellow, he is currently focusing on ultra-low-power and energy-efficient solutions for IoT and wearable markets on the 40-nm NVM and 28-nm FDSOI technologies.

[Chapter 7]

Christoph Zysset

D r. Christoph Zysset is currently working as Application Manager, Computer Controls since 2017, he joined Computer Controls in 2008 as an Application enginner.

Dr Zysset hold MsC in 2008 and PhD 2013 from the swiss federal institute for technology in Zurich his thesis entitled Mechanically Flexible Thin-Film Transistors.

[Chapter 4]